STUDENT UNIT GUIDE

NEW EDITION

OCR A2 Geography Unit F764
Geographical Skills

Michael Raw

PHILIP ALLAN

Philip Allan, an imprint of Hodder Education, an Hachette UK company, Market Place, Deddington, Oxfordshire OX15 0SE

Orders
Bookpoint Ltd, 130 Milton Park, Abingdon, Oxfordshire OX14 4SB
tel: 01235 827827
fax: 01235 400401
e-mail: education@bookpoint.co.uk
Lines are open 9.00 a.m.–5.00 p.m., Monday to Saturday, with a 24-hour message answering service. You can also order through the Philip Allan website: www.philipallan.co.uk

ISBN 978-1-4441-7379-6

First printed 2012
Impression number 5 4 3 2 1
Year 2016 2015 2014 2013 2012

Cover photo: Artman/Fotolia

Typeset by Integra Software Services Pvt. Ltd., Pondicherry, India

Printed in Dubai

Hachette UK's policy is to use papers that are natural, renewable and recyclable products and made from wood grown in sustainable forests. The logging and manufacturing processes are expected to conform to the environmental regulations of the country of origin.

P02136

Contents

Content Guidance

Questions & Answers

Getting the most from this book

Questions & Answers

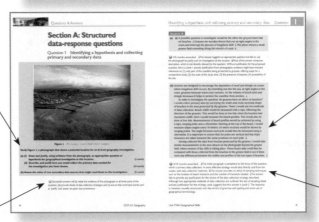

Exam-style questions

Examiner comments on the questions
Tips on what you need to do to gain full marks, indicated by the icon ⓔ.

Sample student answers
Practise the questions, then look at the student answers that follow each set of questions.

Examiner commentary on sample student answers
Find out how many marks each answer would be awarded in the exam and then read the examiner comments (preceded by the icon ⓔ) following each student answer.

About this book

This guide has been written to help you prepare for OCR A2 Geography **Unit F764: Geographical Skills**. The guide is in two parts:

- The **Content Guidance** section defines the specification content and the key themes used to formulate examination questions.
- The **Questions & Answers** section explains the scheme of assessment and outlines the skills required to answer structured data-response and extended-writing questions. It provides seven specimen examination questions (four structured data-response questions for Section A and three extended-writing questions for Section B). Each question has two student answers, ranging in quality from grade A to grade E. Examiner's comments, explaining the marking, are provided at the end of each answer.

Content Guidance

This section provides a summary of the key ideas and content detail needed for A2 Geography Unit F764: Geographical Skills. The content is divided into six main topic areas or stages:

- **Identifying a suitable geographical question or hypothesis for investigation.**
- **Developing a plan and strategy for conducting an investigation.**
- **Collecting and recording data appropriate to a geographical question or hypothesis.**
- **Presenting data in geographical investigations.**
- **Analysing and interpreting data in geographical investigations.**
- **Summarising the findings and evaluating the methodology of an investigation.**

When you revise content it is important to use a framework that reflects how the examiners might test your knowledge and understanding. Therefore, in addition to the key ideas and content detail, this section provides key questions and answers for each topic.

You should study the questions and answers carefully and organise your revision around them. Focusing on the key questions in your revision programme and adding detail of your own to the answers should give you a head start in the final examination.

It is essential that you learn the correct terminology used in the answers to the key questions, particularly the words in **bold** type. You must be confident in applying these terms accurately in your exam answers.

Stage 1: Identifying a suitable geographical question or hypothesis for investigation

Key ideas	Content
A successful geographical investigation depends on identifying a clear geographical question that is practical in research terms.	The questions or hypotheses should be: • at a suitable scale • capable of research • clearly defined and of a clear geographical nature • based upon wider geographical theories, ideas, concepts and processes

How does scale influence the choice of a suitable question or hypothesis?

The success of any geographical investigation is strongly influenced by **scale**. For instance, an investigation of downstream changes in bedload size on a river will depend on the geographical scale chosen for the study. A small-scale investigation over just 3 or 4 km is unlikely to show any significant reduction in bedload size. This is because (a) the processes of **abrasion**, **attrition** and **solution** occur slowly and are apparent only over long distances, and (b) other factors have a significant effect on bedload size at this scale, such as inputs of fresh sediment by tributary streams and localised bank erosion. A study of downstream changes in bedload size, conducted at a larger scale over a distance of 20 or 30 km, is more likely to produce meaningful results.

Scale also has an important influence in vegetation studies. Data on plant cover and biodiversity are normally collected by **quadrat sampling**. However, the choice of quadrat size can have a major influence on outcomes. For example, a 1 m × 1 m quadrat may be too small and fail to record species that are (a) sparsely distributed, and/or (b) geographically clustered.

Examiner tip

Choice of scale is crucial to the success of an investigation. For example, a study of spatial patterns of population density within a market town would use census data for small areal units (i.e. output areas), rather than larger units such as super output areas.

Is it feasible to research the question or hypothesis?

Some questions and **hypotheses** may be impossible to investigate owing to practical problems of measurement (e.g. chemical pollution in rivers), inaccessibility (e.g. research location on private property) or lack of a secondary database. For example, we know that land values and rentals in city centres are related to urban land use, retail location and pedestrian flows. However, there is no easily accessible database for land values and rentals. It is therefore virtually impossible to test a hypothesis such as: *rental values influence footfall in the CBD*. Similarly an investigation into rates of coastal erosion may be hampered by the unavailability of large-scale historic maps.

Other problems also arise when using historical data. In theory it should be possible to investigate changes in population density, ethnic origin and household size between 1971 and 2001 for a suburb in a British town or city. The census of population would be the obvious data source. Unfortunately, the two census years are not comparable owing to changes in census geography (i.e. the **areal units** used to aggregate census data). The old enumeration districts and enumeration areas of the 1971 to 1991 censuses were replaced by the new output and super output areas in 2001, with very different boundaries, thus destroying any continuity.

What is a 'clearly defined hypothesis of a geographical nature?'

A hypothesis is a statement the accuracy of which can be tested using scientific methodology. Geographical hypotheses usually have either a spatial dimension or an overt emphasis on people–environment relationships. In geographical investigation, hypotheses fall into two main types: those that focus on spatial or non-spatial **differences**; and those based on **relationships** between variables. For example, an investigation into the effect of altitude on arable land in a region could be approached through either a hypothesis of difference or a hypothesis of relationship. A possible hypothesis of difference is: *arable land is more extensive below 200 metres than above 200 metres*. An alternative relationship type hypothesis is: *arable land decreases in importance with increasing altitude*.

Clearly defined hypotheses show the direction of differences or relationships. For example, the hypothesis: *shingle beaches are steeper than sand beaches* is more clearly defined than the alternative: *the gradients of shingle beaches differ from the gradients of sand beaches* (see Figure 1).

Michael Raw

Figure 1 Beach gradients: sand and shingle, North Norfolk

The geographical connection must be obvious in hypotheses formulated for geographical enquiry. For instance, a study of shopping patterns in a market centre might consider the frequency of shopping and the type of goods and services purchased. As a result, the following hypothesis might be formulated: *Most shoppers purchase goods and services at least once a week.*

However, this hypothesis is not intrinsically geographical because it lacks a direct connection with either location or people–environment relationships. By adding a reference to the origin of shoppers, we can modify the hypothesis to give it a clear geographical context: *The frequency with which shoppers purchase goods and services is inversely related to journey times from the shoppers' place of residence to the market centre.*

How can questions/hypotheses be based on wider theories, ideas, concepts or processes?

Theory is a set of statements or principles that have been tested and proved valid. In some branches of geography it is possible to formulate questions and hypotheses from a coherent body of theory. For example, central place theory (see Figure 2) provides a conceptual understanding of (a) **settlement hierarchies** and how settlements function as service centres, and (b) how distance and travel costs influence consumers' use of these services.

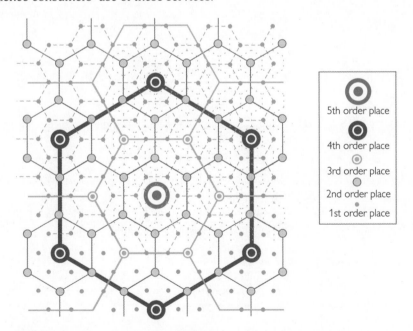

5th order place

4th order place

3rd order place

2nd order place

1st order place

Figure 2 Idealised pattern of settlement according to central place theory

Examiner tip
Exam questions on F764 often require students to relate their knowledge and understanding of geography to the opportunities for geographical investigation presented by stimulus materials such as maps and photographs. Such questions are synoptic and underline the importance of revising all AS and A2 topics prior to the F764 exam.

From this theory we might derive hypotheses such as:

The area served by a central place is influenced by its accessibility, or

The number and range of services provided by a central place depends on the number of people it serves.

Sometimes a topic for geographical enquiry, while lacking formal theory, is supported by a detailed and coherent body of knowledge. It is on the basis of this knowledge that we formulate research questions and hypotheses. For example, from prior knowledge, we know that the development of scree slopes is influenced by gravity, particle size and friction. Thus we are able to formulate a hypothesis such as:

Rock particles on scree slopes are sorted by size.

Knowledge check 2

What is the difference between a hypothesis and a theory?

Summary

- Choice of geographical scale is a crucial decision in identifying a suitable question or hypothesis for investigation.
- A suitable question or hypothesis is one that it is feasible to investigate. Appropriate data must be available and collectable.
- A hypothesis is a statement of difference or relationship between variables whose validity is testable using scientific methodology.
- Hypotheses and geographical questions are derived from either a theory or a coherent body of knowledge on a topic.

Stage 2: Developing a plan and strategy for conducting an investigation

Key ideas	Content
A successful geographical investigation requires careful planning, which balances the need for accuracy against limitations imposed by time, resources and the environment in which the investigation is being conducted.	This stage requires: • the identification of the data needed to examine the question/test the hypothesis • the establishment of appropriate strategies and methods of data collection (including sampling) • an understanding of the limitations imposed by time and resources • an appreciation of the potential risks in undertaking research and the methods of minimising the risk

What data are needed to examine the question/test the hypothesis?

At the outset of an investigation we must identify the data needed to examine a question or test a hypothesis. These data must satisfy three criteria: they must be (a) available at an appropriate scale, (b) available in sufficient quantity and (c) easy to collect. Once identified, the type of data will determine whether collection is by desk research, or fieldwork involving measurement, observation and interview.

Let us assume that a student decides to investigate the geography of poverty in a UK town or city. At an early stage he/she will need to identify a relevant database. In this example, multiple deprivation data for 2004, 2007 and 2010 are available for lower super output areas at the *National Statistics* website. If the study aimed to explain as well as describe the spatial patterns of deprivation, data on ethnicity or unemployment at the same scale would be useful. Again, these data are available from the census of population on the *National Statistics* website. For the purposes of this enquiry, our data satisfy all three criteria, suggesting that the investigation is feasible.

Data collected through fieldwork are more obviously available. Even so, planning is needed to establish that **populations** targeted for study are of sufficient size, are accessible and have the appropriate characteristics. An investigation into the sorting of alluvial sediments on point bars is only practicable on gravel-bed streams transporting coarse bedload. Moreover, it may be decided that sample data must be collected from at least ten point bars, and that all need to be easily accessible. Unless these conditions are met, the investigation may not be feasible.

Before data collection begins, the question of data type needs careful consideration. For instance, will the data be quantitative or qualitative; and which statistical tests (if any) are to be used? Quantitative or numerical data can be used in statistical

> **Knowledge check 3**
>
> What are the likely target populations in investigations into:
>
> • city centre shopping patterns?
> • journeys to work?
> • sediment movement in river channels?
> • river channel efficiency?

Examiner tip
The type of data collected for a geographical enquiry will influence the analytical techniques used in the later stages of investigation. Careful thought is therefore needed at the outset of your investigation to (a) identify the analytical techniques that are to be used and (b) ensure that the type of data collected is suitable for analysis using these techniques.

analysis, but the choice of analytical test will depend on whether data are measured on **ratio**, **interval** or **ordinal** scales. Ratio-scale data are absolute values, which can be used in any test; interval-scale data have no absolute zero (e.g. temperature); and ordinal-scale data are values ranked in order of magnitude. Ratio-scale data, unlike interval and ordinal data, are amenable to all types of statistical analysis. As an example, in a study of beach profiles, the relative position of 10 sites on a beach in relation to mean sea level (i.e. ordinal data — 1, 2...10) might be used. An alternative interval measure for each site would be its height in metres above mean sea level. Ratio-scale data could be collected by measuring the exact position of each site on the beach from mean sea level.

Many studies in human geography that investigate socio-economic and environmental issues are based on qualitative data derived from interviews and questionnaires. In such studies, statistical analysis may be inappropriate. Others, such as investigations of environmental quality and land use, may rely on descriptive or **nominal** categories of data and fieldwork counts that can be converted to numerical data and analysed statistically.

What strategies and methods are needed for successful data collection?

Strategic factors involved in data collection, other than resources, include the timing of the enquiry. **Questionnaire** surveys need to be timed to maximise the number of potential **respondents**. Thus surveys in a shopping centre might take account of the time of day and the advantages/disadvantages of collecting data on weekdays rather than weekends. Similar questionnaire surveys conducted in commuter villages might be timed for a weekend when most people are at home. Coastal studies that involve beach profiling need to consider tide times and are best conducted on an ebb tide or at low tide. Thought should also be given to the location of street interviews. Interviews conducted exclusively at bus stops or in car parks are unlikely to produce accurate and truly representative samples.

Access and the problem of private land ownership is a further consideration. A river, otherwise suitable for a downstream study of changes in channel efficiency, may be inaccessible due to a lack of a public footpath or because the river banks are in private ownership. Similar problems can occur even on common land in the uplands where otherwise attractive locations for river and vegetation studies, are often out of bounds without the permission of the landowner.

Because most statistical populations in geography are huge, fieldwork investigations usually rely on sample data. Decisions must be made on the quantity of data to collect (i.e. sample size) and the selection of sampling strategies that will best ensure accuracy and **representativeness**. In the context of questionnaire surveys the objectives of the investigation will often determine whether street interviews, door-to-door interviews or **postal questionnaires** are most appropriate. The choice of scale also becomes a strategic decision. An enquiry into the spatial pattern of consumer spending in a shopping centre requires a town or city large enough to have at least five or six clearly defined shopping streets/areas within the CBD. Thus, a small urban centre of just 30,000 inhabitants is unlikely to provide a suitable location for such an enquiry.

What are the limitations imposed by time and resources?

Constraints of time and resources impose limits on geographical investigations. As a result, most investigations strike a balance between the amount of data collected and the time and resources expended on data collection. Essentially, this means allocating sufficient time and resources to collect the minimum amount of data needed to answer a question or test a hypothesis successfully.

Some strategic considerations connected to time and resources are practical ones. For instance, how much time will be devoted to data collection? How much potential fieldwork time will be lost travelling (and walking) to and from the fieldwork site? How many students are involved in the data collection process? The last consideration is particularly important in questionnaire surveys, where rejection rates are often high and where each interview may take 3 or 4 minutes. To collect 100 completed questionnaires could take five students all day, whereas 20 students could complete the task in a couple of hours.

What are the potential risks of fieldwork and how can they be minimised?

Geographical fieldwork, both in urban and rural areas, always incurs an element of risk. While we cannot totally eliminate risk, sensible precautions can be taken to minimise it. Thus, before embarking on fieldwork you should do three things. First, identify the potential hazards at the fieldwork site. Second, assess the level of risk presented by each hazard. And finally, devise a strategy or plan for dealing with the hazards. Risk assessment is usually formalised by completing a risk assessment sheet which is then submitted to your teacher/safety officer for approval.

Although the nature of risk varies according to the environment where you are working, a number of general safety guidelines should be observed.
- Make a preliminary visit to the fieldwork site to assess the nature and level of risks.
- Always work in groups. Ideally, groups should comprise at least three people. In the event of an accident one member of the group can assist the injured person while the other can get help (remember that in some environments mobile phone networks may not be available).
- Carry a mobile phone. Leave your phone switched on and make sure that your teacher or a responsible adult has your phone number.

Examiner tip

Any investigation is only as good as the data on which it is based. Common mistakes include: collecting too few data; collecting data in a form that is unsuitable for later analysis; and collecting data that are unrepresentative and unreliable.

Knowledge check 4

What factors need to be taken into account to ensure the personal safety of a researcher during the fieldwork stage of a personal investigation?

- Wear (or carry with you) suitable outdoor clothing (several layers) including waterproofs. For river studies wellies are essential, and for work on slopes and rough terrain, wear walking boots that provide grip and protect your ankles.
- If you are working in a remote area you should carry a torch, a survival bag, a whistle and emergency rations.
- Carry a small first-aid kit and a map of the area (at a suitable scale) at all times.
- Leave precise details of your itinerary with your teacher or a responsible adult, including your time of departure and return.
- Wear safety helmets when working in environments such as cliffed coastlines, screes, boulder fields and steep upland valleys.

Knowledge check 5

What factors need to be taken into account to minimise any risks to the equipment used when collecting data?

Summary

- Data must be collected at an appropriate scale that meets the needs of the study.
- A decision must be made on whether the data required for the study are quantitative, qualitative or both.
- Quantitative data can be measured on ratio, interval and nominal scales. Before data collection begins, it is essential to decide which types of data presentation and statistical tests will be used in the investigation.
- Ratio data give the greatest numerical precision and can be used in all statistical tests.
- Qualitative, non-objective data, based on verbal descriptions, opinions and attitudes may be more appropriate than quantitative data for some types of study (e.g. the social, economic and environmental impact of a proposed wind farm on a community).
- The amount of time available for fieldwork, and the number of students involved in data collection, will determine the amount of data that can be collected. At a later stage, the size of the database will strongly influence the outcome of data analysis.
- The timing of data collection can be crucial to the success of an investigation (e.g. shopping and commuter studies). Problems of access to fieldwork sites (e.g. planned shopping malls, rivers, moorlands) must be resolved before data collection begins.
- Fieldwork is potentially hazardous. Prior to going into the field the potential risks and their level of severity must be defined (ideally by making a preliminary visit) and strategies devised to minimise them.

Stage 3: Collecting and recording data appropriate to a geographical question or hypothesis

Key ideas	Content
A successful geographical investigation is based on thorough methods of data collection and recording, which consider accuracy and reliability in relation to the data being collected.	This stage requires: • the use of primary and secondary data appropriate to the question • a description and explanation of different ways of collecting/recording data • an awareness of the need for accuracy and reliability before, during and after the process of data collection

What are primary and secondary data?

Two types of data are used in geographical investigation: primary and secondary. **Primary data** are original data, which have not previously been collected or processed. All data collected first-hand through fieldwork are primary data. Documents can also be a source of primary data. Trade directories provide information on retailing (e.g. types, number of outlets) which is useful in studies of central place hierarchies and retailing within towns and cities. Similarly, parish registers provide a wealth of primary data on baptisms, marriages and burials, which allow the reconstruction of population trends in the UK before the introduction of civil registration in 1837.

Secondary data comprise information that is available in published documents such as textbooks, articles, maps, charts and diagrams, as well as specific descriptive and analytical techniques and the formulae for their calculation. Today, many of these data are available online. Unlike primary data, secondary data have been processed, ordered and analysed before publication. You would normally expect to acknowledge the source of any secondary data used in an investigation.

Examiner tip

Geographical investigations for F764 must be based around primary fieldwork or primary documentary data. Secondary data will often be used as a context and support for primary data.

What are the different ways of collecting and recording primary fieldwork data?

Examiner tip

Students need to be aware of the pitfalls of questionnaire surveys and questionnaire design. At the outset it is important to define the target population (e.g. shoppers, residents, visitors etc.). Questions must be short and unambiguous, preferably 'closed' rather than 'open', free from bias, and avoid sensitive topics such as income, religion, respondent's age and address. Well-planned questionnaires are both easier to execute and more likely to get a positive response from interviewees.

Primary data collection relies on observation, measurement and interviews. Observation and the systematic recording of data is the simplest data collection technique. **Field sketches** of geographical features, complemented with labels and annotations, rely on observation. Data on traffic flows, land use, shop types, environmental assessment, plant species counts and many other topics are also sourced through observation.

Fieldwork in physical geography invariably requires data collection through measurement. Simple surveying skills generate information on slope profiles and beach gradients; instruments such as anemometers, thermometers and flow meters are used to measure wind speeds, temperatures and water flow velocity respectively; soil pH is measured using a BDH soil testing kit; and callipers are used to measure the long, short and median axes of rock particles. These measurements generate **quantitative data** that can be used to test hypotheses statistically.

Primary data that relate to people's attitudes, behaviour and personal characteristics can only be accessed through questionnaires or interviews. There are two approaches to questionnaire-type surveys. First, those completed by a respondent at home or in the office, where the researcher delivers the questionnaires either by hand or by post. And second, those where the researcher is present to ask the questions and record the responses. These interviews are most often conducted in public spaces (e.g. a shopping street), though occasionally doorstep interviews are undertaken.

What can be done to ensure the accuracy and reliability of data?

Accurate and reliable data are essential to a successful geographical enquiry. Confidence in the outcome of an investigation is undermined where doubts exist concerning the quality of data. The challenge is to acquire objective sample data that accurately represent the statistical population. Three objective sampling techniques are available: **random** sampling; **systematic** sampling; and **stratified** sampling.

Random sampling assumes that every item in the population has an equal chance of inclusion in the sample. This is achieved by using random numbers, generated either from a calculator or from random number tables. Random sampling is often used where a listing of the population exists (i.e. a sampling frame), allowing samples to be selected by random numbers. For instance, households for inclusion in a postal questionnaire could be selected from the electoral register using random numbers.

Systematic sampling offers an alternative to random sampling. Faster and simpler, it involves selecting samples at fixed intervals from a population, though at the outset the sample interval (i.e. every nth item) is determined randomly. Systematic sampling is a more efficient sampling method for street interviews than random sampling. Whereas systematic sampling requires a single random number (to determine the sampling interval), random sampling involves generating a new random number for each potential interviewee.

Both random and systematic sampling strategies are often combined with stratification. Stratification recognises that many statistical populations are not homogeneous, but comprise numerous sub-groups. This is particularly the case with human populations, which are differentiated in terms of age, gender, ethnicity, income, employment, location and so on. Stratified samples are made up of sub-groups, which are represented in the same proportion as in the population. Meanwhile, selection of the samples from the sub-groups can use either random or systematic strategies.

Many geographical investigations focus on the locational characteristics of statistical populations. In these investigations geographers often use **spatial sampling** strategies. Spatial samples are selected from either **points**, **areas** (quadrats) or **transects** (lines) using random, systematic and stratified methods. An enquiry into sediment sorting on a point bar would rely on spatial sampling (see Figure 3). First, a line of transect would be selected randomly and at right angles from the water's edge to the outermost part of the bar. Second, metre **quadrats** (squares) would be located systematically at locations at the start and end, and at equal intervals along the transect. Finally, a sample of say 50 sediment particles at each location would be selected from within each quadrat and their median axes measured and recorded.

Examiner tip

The selection of a sampling strategy should consider a number of criteria such as: the ease with which data can be collected; access to the data collection site; the resources (time and labour) available for sampling; the availability or otherwise of a sampling frame; whether the target population is homogeneous or heterogeneous; the extent to which a sampling strategy produces sample data that represent accurately the population as a whole.

Knowledge check 6

With reference to geographical examples, describe three types of spatial sampling.

Michael Raw

Figure 3 Spatial sampling of sediment on a point bar

Summary

- Primary data are original data collected through either fieldwork or deskwork.
- Secondary data comprise information that has previously been processed, ordered and analysed, and often published in articles, textbooks and online.
- The aim of sampling is to produce sample data that accurately represent a population.
- Accuracy and reliability in primary data collection are achieved by objective, scientific sampling methods.
- There are three general sampling strategies: random, systematic and stratified.
- Stratified sampling is used where populations comprise a number of sub-groups. Samples are selected from these sub-groups either randomly or systematically.
- Spatial sampling, based on points, areas and transects, is used where the locational characteristics of a population are of particular interest to an investigation.

Stage 4: Presenting data in geographical investigations

Key ideas	Content
A successful geographical investigation involves the selection of techniques that are appropriate to the data collected and their presentation to a high standard.	This stage requires: • the use of appropriate techniques (including maps, diagrams, annotated photographs and charts) to present the data collected • the logical organisation of the presented material in relation to the analysis • presentation of the material to a high standard relevant to the question/hypothesis posed

What mapping techniques are used to present geographical data?

Statistical maps are the most effective way to present spatial data. There are five types of statistical map: **dot**, **choropleth**, **isopleth**, **proportional symbol** and **flow** maps. Their effectiveness can be assessed by the following criteria:

- accuracy and detail
- visual impact and clarity
- general description of spatial patterns and trends
- ease with which statistical information can be retrieved

Knowledge check 7

What criteria would you use to evaluate the effectiveness of a statistical map?

To an extent, all statistical maps are a compromise because they represent a trade-off between accuracy and detail, and visual impact and clarity.

Dot maps show the location of a given quantity, such as number of people or hectares of wheat, by a dot of constant size (see page 53). They often provide an excellent visual impression of geographical distributions and, unlike choropleth maps, are not interrupted by the boundaries of areal units (e.g. super output areas, postcode sectors). However, dot maps have disadvantages. Apart from being time-consuming to construct, dot placement is highly subjective and can give a misleading impression of accuracy. In high-density areas, where dots begin to merge, data recovery may be difficult; while in low-density areas, they convey little idea of the actual distribution.

Choropleth maps show the spatial distribution of data by areas. Also called proportional shading maps, they represent geographical differences between areas by colour or intensity of shading. Unlike dot maps, choropleth maps rely on standardised data independent of the size of areal units. Thus choropleth mapping is used to show spatial patterns of population density and percentage population change in an area, rather than population totals or absolute population change.

Choropleth maps are easy to draw and have great utility because statistics are usually available for standard areal units such as local authorities, wards and super output

areas. However, choropleth maps also have shortcomings: they tell us nothing about the internal distribution of values within areal units (a particular problem where units are large); substantial variations in the size of areal units often dominate the appearance of the map; and the boundaries of areal units produce abrupt discontinuities in spatial patterns that are not present in reality.

Proportional symbol maps represent absolute values such as population counts and crop areas, with circles, squares and triangles (see page 53). The area of each symbol is proportional to the value it represents. While proportional symbol maps give a clear and immediate visual impression of geographical distributions they have a number of drawbacks. Placement of symbols is often arbitrary; the maps provide no information on distribution of values within areal units; and overlapping symbols can make interpretation difficult in high-density areas.

Isopleth maps show spatial distributions with continuous lines (known as **isopleths** or **isolines)** (see Figure 4). Familiar examples of isopleth maps include topographic maps (contours), temperature maps (isotherms) and rainfall maps (isohyets). Isopleth maps are the most appropriate method for showing spatial distributions which have a continuous surface. However, because these maps are extrapolated from point values, their construction often involves a large degree of generalisation. Over large areas, where values are smoothly distributed, this is not a problem. However, at a local scale and with irregular distributions (e.g. land values in cities) the accuracy of isopleth maps is more questionable and their interpretation can be difficult.

> **Knowledge check 8**
>
> Comment on the effectiveness of isopleths in mapping the distribution of human populations.

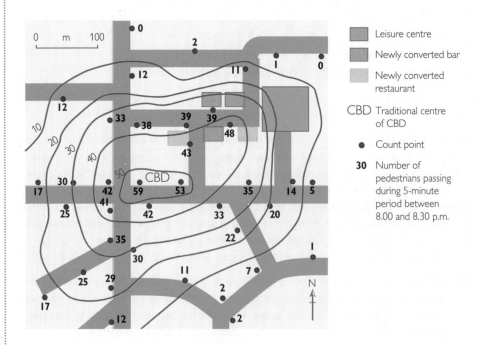

Figure 4 An example of an isoline map: pedestrian densities in a town centre

Flow maps show the movement of people, vehicular traffic, goods and information over space. Movements are represented by lines that are proportional in thickness to flow volumes. Flow paths can be either non-routed (i.e. usually straight lines) or routed, where flows follow actual pathways along streets and other transport networks. Non-routed flow maps can be excessively generalised and have limited value for small-scale movements, such as pedestrian flows in city centres. Routed flow maps, based on point values (e.g. traffic counts), can also mislead: abrupt changes of flow often occur at count points and may not correspond to actual changes on the ground.

What graphical techniques are used to present geographical data?

Data presented in charts, rather than data arrays and tables, have two main advantages. First, they provide a convenient statistical summary of geographical data, and second, through their visual impact, they assist assimilation of data and the identification of patterns and trends.

Charting data is an important stage in geographical investigation and is essential to the preparation of data for hypothesis testing. The most widely used charts in geographical investigation are **histograms**, **bar charts**, **pie charts**, **line charts** and **scatter charts**.

Histograms are a type of bar chart where the frequency distribution of data is presented in classes. The stepped appearance of histograms is due to the arbitrary choice of class intervals. Thus the irregular shape of a histogram is often smoothed-out and replaced with a **frequency distribution curve** (see Figure 5). Such curves may be either **normal**, with a symmetrical shape, or asymmetrical (i.e. **skewed**). In geographical populations we often find that frequency distributions are **positively skewed**, with the 'tail' of the distribution extending to the right (e.g. slope angles, sediment size). **Negatively skewed** populations, where the 'tail' extends to the left, are less common.

Bar charts comprise a series of rectangles proportional in length and area to the values they represent. They are appropriate for presenting data that relate to **discrete** places, groups or time units (e.g. age-sex population structures, inter-censal population changes for local authority areas). **Stacked bar charts** represent two or more data sets by subdividing the bars. They may show absolute values (individual bars will vary in size) or proportional values such as percentages (individual bars are the same size).

Where data are **continuous** rather than discrete (e.g. a hydrograph showing river discharge over a month or year), **line charts** are a more appropriate descriptor than bar charts.

Examiner tip

You must have valid reasons for choosing to use a particular type of statistical map. Justifications for your choice might include: the type of data available; whether the data represent areas or points; the accuracy and degree of spatial detail required; and the ease with which a map can be drawn or numerical data recovered from the map.

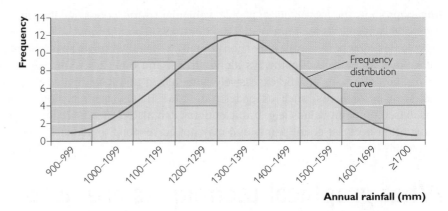

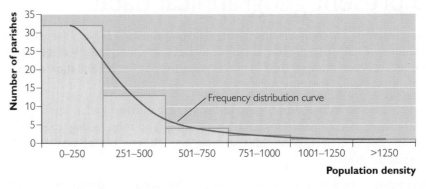

Figure 5 Two types of frequency distribution curve

Pie charts are circular graphs, subdivided into segments to show sub-groups in a population. The total population is represented by the area of the chart. The first segment of the chart normally starts at zero degrees, and for clarity, the number of segments is usually limited to a maximum of seven or eight.

Scatter charts are used to plot two variables, x and y. Variable x, the **independent variable**, which causes change in y, the **dependent variable**, is plotted on the horizontal axis. The dependent variable (y) occupies the vertical axis. Scatter charts provide a visual impression of the relationship between variables: the closer the scatter of points to a straight line, the stronger the relationship. If the points on a scatter chart trend from bottom left to top right, the relationship is **positive** (i.e. as x increases there is a proportional increase in y). An **inverse** or **negative** relationship trends from top left to bottom right and shows that an increase in x produces a corresponding decrease in y.

How should data be organised and presented in relation to the analysis?

Data are presented as tables, maps and charts as the initial stage of data analysis. At this stage data are organised and grouped according to the questions/hypotheses under investigation. For example, an enquiry into household movements originating within a city suburb might be sequenced by: (1) mapping the location of new residences, (2) presenting the data first as a frequency table showing distances moved, (3) presenting the same data as a histogram. The process of data presentation is an essential prelude to more rigorous analysis using statistical methods. It also provides guidance on whether further analysis (or otherwise) is worthwhile.

Knowledge check 9

Assess the relative merits of choropleth and isopleth maps for showing the distribution of population in a medium-sized city.

Summary

- Data presentation in the form of tables, charts and maps, showing patterns and trends, is an essential precursor to data analysis.

- Statistical mapping techniques, such as dot, choropleth, isopleth, proportional symbols and flow mapping, are used to represent spatial distributions.

- Graphical techniques, such as histograms, and bar, line, pie and scatter charts, generalise data sets, storing information and showing patterns and trends.

- The choice of a specific mapping or graphical technique to represent a data set will be influenced by factors such as scale, types of data (e.g. spatially discrete or continuous, absolute or standardised), the amount of detail required, and the ease or difficulty of construction.

- Any assessment of a specific mapping or graphical technique should take into account the clarity, accuracy, effectiveness and detail of the representation, as well as the amount of numerical data that can be stored and extracted from the map or chart.

Stage 5: Analysing and interpreting data in geographical investigations

Key ideas	Content
A successful geographical investigation involves a variety of analytical approaches ranging from the descriptive to detailed analysis.	This stage requires: • the description of findings shown by data presentation • the analysis of data using statistical techniques • the interpretation of results in relation to the original question/hypothesis posed • explanations of patterns found and any anomalous results

How is the information shown by data presentation methods described?

As a preliminary to data analysis, the information presented in charts and maps is first described. Accurate and comprehensive data description comprises (a) a description of the main patterns and trends, (b) exemplification of the patterns and trends using specific data from the relevant charts or maps, and (c) references to exceptions or **anomalies** which deviate from the main patterns and trends. Descriptions should not be over-generalised, but at the same time they must not be so detailed that they fail to identify the main patterns and trends.

How are descriptive statistical techniques used in data analysis?

Descriptive statistical techniques comprise measures of central tendency and dispersion. They are used to describe or summarise data sets.

We often need to represent a data set with a single value. To do this we choose a middle value around which the data cluster, known as a measure of **central tendency**. Three standard measures of central tendency are used: the **arithmetic mean** or average; the **median**; and the **principal mode**.

The **arithmetic mean** is the sum of all values in a data set divided by the number of values. Though mathematically based and the most widely used measure of central tendency, the mean needs to be interpreted with care. This is because it weights each value according to its magnitude. Thus a handful of very large values, which may not be typical of the rest of the data, can inflate the value of the mean. The mean

Knowledge check 10

Under what circumstances would you use the median, rather than the mean, to summarise a data set?

OCR A2 Geography

is most accurate when a data set approximates a normal frequency distribution and has a narrow range.

The **median** is the middle value in a data set arranged in order of magnitude and is therefore unaffected by extreme values. When distributions are highly skewed, the median provides a more representative summary than the mean.

The **principal mode** is the class in a histogram (or frequency table) that contains the most values. Although it provides a good general description of a data set, it has weaknesses. Unlike the mean and the median the principal mode is a range of values rather than a single value; some distributions have more than one mode; and the principal mode's value ultimately depends on the arbitrary choice of class intervals used in a histogram.

Measures of central tendency tell us nothing about the dispersion of values in a data set. It is important to have a summary measure of dispersion because very different data sets can yield similar values for the mean, median and mode. The main measures of dispersion are the **range**, **inter-quartile range** and **standard deviation**.

The **range** is the difference between the highest and lowest values in a data set. It is the simplest measure of dispersion and is often used in weather and climate statistics. The **inter-quartile range**, used in conjunction with the median, and based on half the values in a data set, is more representative than the range. Data are arranged in rank order and divided into four equal parts or **quartiles**. The boundary separating the upper 25% of values is the **upper quartile**; the **lower quartile** is the boundary defining the lowest 25% of values. The inter-quartile range is the difference between the upper and lower quartiles.

The **standard deviation** (or root mean square deviation — which describes its calculation) is the most accurate measure of dispersion, and is used alongside the mean. All values in a data set are used in its calculation. Because the standard deviation is strongly influenced by the magnitude of the mean, it is often standardised by expressing it as a percentage of the mean. The resulting statistic, which allows comparison of dispersion in different data sets, is known as the **coefficient of variation**.

> **Examiner tip**
> Although it is useful to know how to calculate measures of central tendency and dispersion, more important is a critical appreciation of their strengths and weaknesses and when it is appropriate to use each of them.

How are inferential statistical techniques used in data analysis?

Statistical techniques are used to test hypotheses objectively. Whatever the results of an investigation, there is always a possibility that outcomes could be due to chance. This is because most geographical enquiries are based on samples rather than the statistical population. Statistical techniques allow us to assess the probability that the results of an enquiry are due to chance. If this probability is small, we can accept the results with a high level of confidence.

Most hypotheses are formulated in one of two ways: they either focus on **differences** between data sets or on the **relationships** between them. The nature of any hypothesis will determine the choice of statistical test. Unit F764

specifies three inferential tests. Hypotheses of difference are analysed either by the **Mann-Whitney U** test or the **Chi-squared** test. Hypotheses that test relationships use correlation methods such as **Spearman's Rank** correlation coefficient.

The **Mann-Whitney U** test analyses the statistical significance of the difference between two data sets. It could be used, for example, to test the hypothesis that the cross-sectional shape of a river channel is more efficient in straight than in meandering channels. Data to test this hypothesis might comprise sample values of channel efficiency (e.g. width to depth ratio) for straight channels, and similar sample measurements for meandering channels (see Table 1). The U test is based on the rank order of all values (i.e. 17 in Table 1) in the data set. The outcome is statistically significant if the U value is less than the critical value at the 95% level in U tables. The U test's main limitation is that it can only be used to analyse two data sets. However, it can be used for relatively small samples and, being a non-parametric test, it is not conditional on the statistical population having a normal frequency distribution.

Table 1   Channel efficiency in straight and meandering channels: Marshaw Wyre, Lancashire

Straight channels	Meandering channels
6.57	29.73
14.83	14.85
11.76	28.35
6.052	28.43
10.40	14.21
9.88	23.24
14.79	36.29
9.44	
10.37	
12.77	

Correlation tests such as **Spearman's Rank** and **Pearson's Product Moment** measure the strength of the relationship between two variables and the statistical significance of the relationship. The outcome of correlation analysis is a correlation coefficient which ranges in value from −1 to +1. A coefficient of −1 describes a perfect **negative** or **inverse correlation** while +1 indicates a perfect **positive correlation** (see Figure 6). The closer the coefficient is to zero, the weaker the relationship.

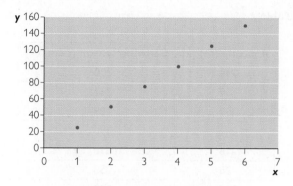

(a) Perfect positive correlation: +1

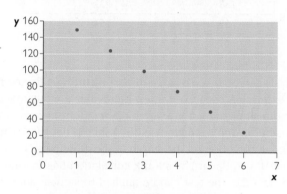

(b) Perfect inverse correlation: –1

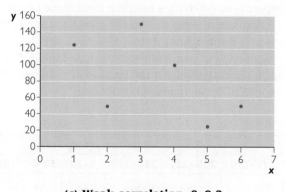

(c) Weak correlation: 0–0.2

Figure 6   Scatter charts and correlation

Correlation in A-level specifications deals only with situations that involve two variables (*x* and *y*). Normally we assume a relationship of causality, with *x* inducing change in *y*. An example would be the relationship between altitude (*x*) and the length of the growing season (*y*) (see Table 2). Having established the strength of a relationship with a correlation coefficient, we need to know the likelihood that the correlation has occurred by chance. To do this we again refer to statistical tables.

Knowledge check 12

For the correlation hypotheses listed in Knowledge check 11 (p. 26), identify the *x* and *y* variables.

Table 2 Average altitude and length of growing season: northwest England and north Wales

Altitude (m) (x)	Growing season (days) (y)
214	200
109	237
315	189
341	190
116	238
28	258
287	209
138	232
65	256
61	277

Correlation analysis can be taken a stage further by fitting a **linear regression** line to the data set. Regression does two things. It provides a mathematical summary or model of the relationship, and it allows prediction of y from known values of x. The latter could be useful in the example used in Table 2, because while altitude data are easy to obtain, information on the length of the growing season is not so widely available. Like the U test, Spearman's Rank correlation has the advantage of being a non-parametric test (i.e. the test can be applied to sample data regardless of the frequency distribution characteristics of the parent population).

Chi-squared compares the differences between an observed frequency distribution and an expected distribution generated under random conditions. Data must be grouped in classes in a **contingency table**. Moreover, the data must be absolute values, not percentages or other standardised forms (see Table 3). The main advantage of the Chi-squared test is its flexibility. It can be used to analyse differences of two or more distributions, and as an alternative to correlation.

Table 3 Orientation of long axes of ploughing blocks below Ilkley Crag, West Yorkshire

	Alignment (degrees)						
	270–299	300–329	330–359	0–29	30–59	60–89	90–119
Number of blocks	12	18	38	53	22	7	1

Examiner tip
Geographical investigations should, at the outset, define the statistical methods to be used in data analysis. Knowing how data are to be analysed will influence the type and amount of data collected. Planning is essential and any investigation should be viewed as a holistic enterprise, and not just a series of discrete stages (e.g. data collection, data presentation, data analysis).

For example, a study of scree particles on a talus slope could measure distance downslope every 10 metres, and from a sample of particles establish the mean particle size at each location. Analysis could then be done by correlation. Alternatively, five or six broad distance zones could be identified on the slope, and the frequency distribution of particles for each zone established. These data would then be in an appropriate form for Chi-squared analysis. The outcome of this test is a Chi-squared statistic, where the statistical significance is obtained from Chi-squared tables. The larger the value of the Chi-squared statistic, the greater the probability that the result is not due to chance. Chi-squared, like the U test and Spearman's Rank correlation, also has the advantage of being a non-parametric test.

How are results interpreted in relation to the original question/ hypothesis posed?

We interpret the results of geographical investigation by assessing their **statistical significance**. Essentially this means establishing the probability that the results are due to chance. Statistical significance is normally set at the 95% (or 0.05) threshold. At this level, there is only a 5% probability (or 1 chance in 20) that the result has occurred purely by chance. Because we can be confident in the result, we accept our hypothesis. Alternatively, a hypothesis which failed to reach the 95% level would be rejected.

Statistical significance is obtained from statistical tables. Separate tables are used for the U test, Spearman's Rank correlation and Chi-squared. In order to use the tables we need to (a) calculate a value for the statistical test employed (b) define the level of significance required (e.g. 95%, 97.5%, 99%), and (c) determine the number of degrees of freedom i.e. the number of pieces of independent information contained within a particular analysis — usually $n-1$.

> **Knowledge check 13**
>
> Explain the importance of statistical significance in geographical investigations.

> **Knowledge check 14**
>
> Describe a number of situations in geographical investigation when it would be appropriate to use the U test, Chi-squared test and Spearman Rank correlation.

How do we explain the patterns found and any anomalous results?

If the patterns found in an investigation are consistent with the hypothesis being tested, then the hypothesis and hence the theory or logic which underpins it, are verified. Sometimes a broad pattern or trend is apparent, but a few values deviate from the expected. Such anomalies may be explained by the multivariate nature of many geographical features (i.e. their explanation involves more than one causal factor) or specific problems of data collection such as sampling accuracy.

Anomalies of this kind are exceptional, but do not necessarily invalidate a hypothesis. They do, however, demand explanation. Figure 7 shows the results of an investigation into the average discharge and drainage basin area for a number of rivers in northern England. There is a clear pattern, with discharge positively correlated with drainage basin area. Even so, one river is clearly anomalous and deviates significantly from the trend. This anomaly — the Gypsy Race in east Yorkshire — is explained by the relatively low rainfall and high evapotranspiration in this part of northern England. Other factors which might account for anomalies include water abstraction, reservoirs (which increase evaporation losses) and vegetation cover (e.g. rates of interception are higher in forested catchments).

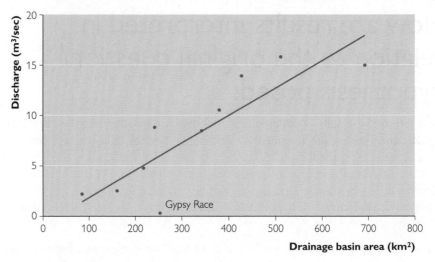

Figure 7 Scatter chart and regression line: drainage basin area and average discharge in northern England

Summary

- Descriptions of data presented as tables, charts and maps should: (a) identify the main patterns and trends, (b) highlight any anomalies and (c) illustrate patterns, trends and anomalies by referring to specific values and examples.
- Measures of central tendency summarise data sets with a single, middle value. The mean is the most useful measure of central tendency. However, where distributions are skewed the median may be more representative than the mean.
- The distribution of data around a central value is summarised by measures of dispersion, such as the range, the inter-quartile range and the standard deviation.
- Inferential statistical tests are most often used to analyse sample data drawn from a much larger statistical population. These tests tell us the probability that results based on sample data have occurred by chance.
- The U test analyses the differences in two data sets; Spearman's Rank correlation measures the association of (or relationship between) two variables; while Chi-squared is used to test both hypotheses of difference and hypotheses of association.

Stage 6: Summarising the findings and evaluating the methodology of an investigation

Key ideas	Content
A successful geographical investigation involves a clear summary of its findings and an evaluation of its methodology.	This stage requires: • the use of evidence presented in previous sections to provide a clear conclusion which relates back to the original question/hypothesis posed • an evaluation of the extent to which a study supports or otherwise the general theories, ideas and concepts being studied • an evaluation of the limitations of the study in terms of methods used and data collected, and suggestions for possible improvements

How does an investigation provide a clear conclusion which relates back to the original question/hypothesis posed?

At the outset of an investigation we establish its aims and objectives. For the most part these comprise a series of questions and/or hypotheses. In the final stage of enquiry we formulate a conclusion which (a) sets out our main findings, and (b) reflects on the extent to which the aims and objectives of the study have been achieved.

In geographical investigations conclusions are often equivocal and rarely clear-cut. This is not an admission of failure. It arises because geographical situations are invariably complex, and data collection cannot be controlled with complete accuracy. For example, a study of the influence of geography on rates of obesity in two suburban neighbourhoods might prove inconclusive because obesity is not just influenced by access to gyms, parks and health clubs. It is also affected by income, personal preferences, perception of opportunities, age, ethnicity and so on. Moreover, collecting representative sample data on lifestyles is extremely challenging.

To what extent does a study support general theories, ideas and concepts?

Questions and hypotheses are usually derived from a general body of theory, ideas or concepts. It is useful to place a geographical investigation within its wider context and

Examiner tip

Conclusions should provide an honest assessment of both the strengths and weaknesses of an investigation. Awareness of a study's limitations will gain credit from the examiners. The outcome of a study is far less important than showing understanding of six stages of geographical investigation and applying them accurately and relevantly to a topic of your choice.

comment on the extent to which its findings are consistent with previous enquiries. Studies of step-pool sequences on boulder-bed streams in California have shown a regularity of spacing, and a strong relationship between the ratio of channel width to distance between steps, and channel gradient. A personal investigation of step-pools might, in its conclusion, compare its findings to other similar studies. In particular it might draw attention to local factors such as geology, bedload calibre and relief that have contributed to different outcomes.

What are the limitations of a study in terms of methods used and data collected, and what possible improvements could be made?

It is almost certain that any geographical investigation at A-level will encounter limitations in terms of methodology. The most obvious limitations concern the choice and execution of the sampling strategy, and the amount and quality of data collected. In questionnaire surveys, achieving a truly representative sample, based either on random or systematic methods, is extremely difficult. Rejection rates for street interviews are high and researchers often resort to interviewing anyone willing to respond. The result is a sample that is unrepresentative and inaccurate. From the outset this defective sample will undermine the validity of the study. Even where a random or systematic approach is adhered to, without some stratification, sample interview data are usually biased. It is well known that some groups, such as older adults and women, are more likely to respond to questionnaire surveys than others. These problems underline the need to (a) adopt a consistent random or systematic sampling method, and (b) where appropriate, ensure that samples are stratified.

Insufficient sample data imposes further limitations on many geographical investigations. An inadequate database compromises statistical analysis, giving little confidence in the reliability of results. This problem can be tackled at the planning stage by identifying the statistical test used for analysis and the minimum size of sample needed to obtain statistically significant results. Where there is doubt, it is preferable to collect larger, rather than smaller samples.

Knowledge check 15

An A-level geographical investigation into shopping in a market town of 25,000 people failed to show that the average journey time of shoppers using a large Tesco supermarket was longer than for shoppers using the smaller Aldi store. Suggest possible methodological reasons that might explain this unexpected outcome.

Summary

- The concluding stage of an investigation will (a) state the investigation's main findings and (b) consider the extent to which it has achieved its aims and objectives.
- The conclusion should include some discussion of the consistency of the investigation's findings compared with existing theories, models, concepts and similar studies.

- The conclusion will discuss the limitations of the investigation, with particular reference to methodology (e.g. data collection strategies), including the quality, amount and appropriateness of the primary database.

Questions & Answers

Assessment

F764 Geographical Skills is one of the two units that make up the A2 specification. It is worth 80 uniform marks and accounts for 40% of the A2, and 20% of the A-level specification weighting. The other A2 unit is Global Issues (see Table 1).

Table 1   A2 Geography: scheme of assessment

Unit number	Unit name	Unit exam length	Raw marks	Uniform marks	A2 weighting (%)
F763	Global Issues	2.5 hours	75	120	60
F764	Geographical Skills	1.5 hours	60	80	40

Unit F764 covers the six main stages of geographical enquiry:

1 Identifying a suitable geographical question or hypothesis.
2 Developing a plan or strategy for conducting the investigation.
3 Collecting and recording appropriate data.
4 Presenting the data collected in appropriate forms.
5 Analysing and interpreting the data.
6 Summarising the main findings and evaluating the investigation.

The exam paper is in two parts — Section A and Section B. Section A requires you to answer **one** structured data-response question from a choice of three. In Section B you must answer both of the extended-writing (i.e. essay) questions. (Note that three extended writing questions have been provided in the Questions and Answers section of this guide.)

Structured data-response questions

These questions, based on stimulus materials such as sketch maps, diagrams, charts, newspaper extracts and photographs, are divided into three sub-sections (a, b and c), worth 5, 10 and 5 marks respectively. Each question should take around 30 minutes to complete. The questions focus on your understanding of the six stages of geographical enquiry and are generic in nature. Thus you are not expected to have carried out any investigation or specific study of the topics in question. There is also some emphasis on new technologies in geographical research, including computer skills and the application of Geographical Information Systems (GIS).

Extended-writing questions

The extended-writing questions in Section B focus on the skills and techniques you have used in your personal fieldwork-based/research-based investigations. They will

typically relate to analysis, interpretation, evaluation and the drawing of conclusions from geographical enquiries. You should give yourself approximately 30 minutes to answer each question in Section B.

Mark scheme criteria

Examination answers are assessed against three criteria or assessment objectives (AOs). The three AOs for A2 Geography are:

1 **Demonstrate knowledge and understanding** of the specification content, concepts and processes.
2 **Analyse, interpret and evaluate** geographical information, issues, viewpoints and apply them in unfamiliar contexts.
3 **Investigate, conclude and communicate,** by selecting and using a variety of methods, skills and techniques to investigate questions and issues, reach conclusions and communicate findings.

It is worth studying these criteria carefully because they tell you how your examination answers will be judged. Meanwhile the section on examination skills below explains how assessment objectives are used in the mark scheme to assess your answers. Table 2 shows the weighting given to each AO.

Table 2 Assessment objective weightings in A2 Geography

Unit		% of A2			
		AO1	**AO2**	**AO3**	**Total %**
A2 Unit F763	Global Issues	20	30	10	60
A2 Unit F764	Geographical Skills	10	10	20	40
Total %		**30**	**40**	**30**	**100**

Examination skills

Success in A2 Geography requires not only good knowledge and understanding of the specification content but also an effective exam technique.

To acquire a solid knowledge base that is both relevant to the specification content and to the style of the examination you should structure your revision around the key ideas and questions in the **Content Guidance** section of this unit guide. This structure will help focus your learning on the areas most frequently targeted by examiners.

In addition to revising the specification content, you must spend time developing and refining your examination technique. The most obvious indication of poor technique is a failure to apply knowledge and understanding appropriately to a question, and this invariably results in a heavy loss of marks. While marks are available for showing knowledge and understanding, accessing the higher attainment levels in the more demanding questions requires you to apply understanding relevantly and in unfamiliar contexts.

Answering structured data-response questions

The structured data-response questions in Unit F764 are based on a range of stimulus materials, including charts, maps, satellite images, photographs, diagrams and statistical information. Each question is divided into three parts. The first sub-question (a — worth 5 marks) requires you to respond directly to the stimulus material, usually by identifying or describing patterns, trends, spatial differences and other features. It is generally less demanding than the later sub-questions (b) and (c).

Questions (b) and (c) are less specific and test your wider, generic knowledge and understanding of geographical enquiry. In the context of the question you might, for example, be asked to describe and justify an appropriate method of data sampling, data presentation or statistical analysis. This type of question, which makes greater intellectual demands than straightforward description, is worth 10 marks. The third sub-question, worth 5 marks, requires less depth than (b) but is often more open-ended and ideas-based. You can prepare for this question by drawing on your experience of fieldwork and research, and by showing an awareness that geographical investigation is complex and for many reasons often fails to produce expected outcomes. This may be due to range of factors, such as: poorly formulated hypotheses or questions; inappropriate scales of investigation; problems of collecting representative samples; inadequate sample size; the multivariate nature of geographical phenomena affected by many causal factors; and so on.

Mark schemes for structured data-response questions are levels-based (see Table 3). There are two levels of attainment for 5-mark questions, and three levels for 10-mark questions. Marks are weighted towards the top end. For example, in a 10-mark question, 8+ marks are reserved for level three answers. Having read your answer, the examiner will first decide its level and then allocate it a precise mark within that level.

Table 3 Basic mark scheme descriptors for structured questions

5-mark questions		
Level	Marks	General descriptor
2	4–5	Clear/detailed/accurate description/justification/explanation. Clear reference to stimulus materials.
I	0–3	Limited outline/description/explanation. Little if any reference to stimulus materials.

10-mark questions		
Level	Marks	General descriptor
3	8–10	Clear/accurate/in depth description/justification/explanation. Well structured with accurate grammar and spelling. Good use of appropriate geographical terminology.
2	5–7	Sound detail/accuracy, some justification/explanation. Sound structure but may have some errors in grammar and spelling. Some use of appropriate geographical terminology.
I	0–4	Limited description/detail/justification/explanation. Little structure, and has some errors of grammar and spelling. Little use of appropriate geographical terminology.

When answering structured data-response questions it is important that you follow these guidelines:

- Read through all parts of the question before attempting to answer. This will help you to avoid possible repetition in later answers and allow you to get an overview of how the topic is developed.
- Study the stimulus material carefully.
- Before writing make sure that you understand precisely what each question is asking you to do.
- For the 10-mark question, which requires up to 25 lines of writing, you will need to plan your answer. Make a list of key points and any examples of fieldwork/research investigations that you want to use in your answer.
- Divide your time realistically and adjust the length of your answers to the mark weighting. For example, a 5-mark question is unlikely to require more than 10 lines, while a 10-mark question will require you to write at least twice as much.

Answering extended-writing questions

Section B of Geographical Skills comprises two extended-writing questions, each worth 20 marks. Two questions are set, and you are required to answer both. At first sight this lack of choice appears to be quite demanding. However, it is less of a constraint than you might think because both questions ask you to write about specific fieldwork/research investigations that you have completed during your A-level course. The two essays can be related to any of the six stages of geographical enquiry in the context of your own investigation.

The mark weighting for these questions suggests that you spend around 30 minutes on each answer, including 3–4 minutes of thinking and planning time.

As a rule, extended-writing questions in Section B include basic command words such as *describe*, *how* and *explain*, together with evaluative commands like *assess*, *to what extent...?* and *how important...?* (see page 38).

Mark schemes for the extended-writing questions have three levels of attainment (see Table 4). They are not structured by the assessment objectives. Instead, all the assessment objectives are subsumed within the description of each level. As a result, the mark schemes are a little more variable than in Section A and tend to reflect the nature of the question.

Table 4 Mark scheme descriptors for extended-writing questions

Level	Marks	Descriptor
3	16–20	Clear and detailed description/explanation/evaluation. Cause and effect are clear. Answers are clearly linked to specific fieldwork/research investigations. Answers are well structured. There is accurate use of spelling, grammar and geographical terminology.
2	10–15	Limited detail in relation to description/explanation/evaluation. Some cause and effect attempted. Some linkage to specific fieldwork/research investigations. Sound structure, but with some inaccuracies of spelling and grammar. Some use of appropriate geographical terminology.
1	0–9	Some description/explanation but little, if any, evaluation. No real cause and effect and limited, if any, linkage to actual fieldwork/research investigations. Answer has little structure and some errors of spelling and grammar. Little use of appropriate geographical terminology.

Extended-writing answers to the questions in Section B must be presented within the context of fieldwork or research enquiries that you have undertaken personally. Answers that are largely generalised are unlikely to achieve more than a moderate Level 2 standard.

You can make your answers enquiry-specific in a number of ways. For example, you could provide details of the area studied and the precise nature of the investigation; outline the influence of the local environment in the formulation of hypotheses and sampling strategies; and discuss the unique problems relating to the investigation.

In the context of spatial sampling strategies, it is particularly helpful to the examiner if your written description is supported by a sketch diagram. Examples are shown on pages 59, 61, and 66–67 of the Questions & Answers section.

Successful extended-writing answers should:
- Demonstrate accurate and detailed knowledge and understanding of the topic.
- Support general points by using appropriate exemplification.
- Emphasise discussion and provide some evaluation.
- Be logical and structured, with an introduction and conclusion.
- Use language and geographical terminology clearly and accurately.

Figure 1 shows a typical extended-writing question and indicates how the question provides opportunities to assess knowledge, understanding, evaluation and exemplification.

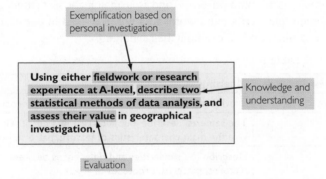

Figure 1 The main features of an extended writing question on geographical skills

Before attempting to answer an extended-writing question it is important to spend a few minutes planning your ideas and defining its content. Your plan should be sufficiently detailed to list the main content of each paragraph and the geographical examples you intend to use to support your answer. An effective answer will have three main components: an introduction, a main body, and a conclusion. These components are illustrated here by referring to the question in Figure 1:
- **Introduction** The introduction should define any key terms used in the question such as 'statistical methods' and should indicate the broad structure of your answer. You should name the two statistical methods (e.g. correlation, U test) then briefly outline the nature of the fieldwork or research investigations you intend to use. It is important that the introduction is kept brief and to the point.
- **Main body** The main body will include the substance of your answer, with detailed knowledge and understanding, exemplification and, where appropriate, evaluation. In this question, the simplest approach would be to deal with each statistical method separately. First you would describe the statistical method (i.e. what it does, the type

of data it uses); then you would show how it was used in the context of a specific geographical enquiry; and finally you would assess its value to the enquiry.

- **Conclusion** The conclusion, like the introduction, should be brief. It should summarise the main points developed in your answer, and in a discursive question would normally include some overall evaluation. In this example your conclusion might emphasise the value of statistical analysis in establishing the statistical significance of outcomes.

Fieldwork and research investigations

All answers to extended-writing questions in Section B require knowledge and understanding of appropriate fieldwork or desk-based research investigations. Aspects of these investigations, such as sampling methods, data collection, data presentation and data analysis, provide crucial and detailed exemplification. Without this, answers become generalised and are likely to achieve Level 2 at best (see Table 4).

Command words and phrases

The command words and phrases in examination questions are crucial because they tell you exactly what you have to do. You must respond accurately to their instructions. For example, the instruction 'describe' is very different from 'explain'. Ignoring command words and phrases is a common error, and is a major cause of underachievement. Table 5 lists some typical command words and phrases used in questions in the OCR A2 Geography examination and explains their meaning.

Table 5 Key command words and command phrases

Command word/phrase	Meaning
Describe	Provide a word picture of a feature, pattern, process or idea.
Outline	The same as 'describe' but requiring less detail. The idea is to give the main characteristics of feature, pattern, process or idea.
Compare	Describe the similarities and differences between two or more features, patterns, processes or ideas.
Examine	Describe and comment on a pattern, process or idea. 'Examine' often refers to ideas or arguments, which demand close scrutiny from different viewpoints.
Justify	Prove or explain the use of; vindicate.
Suggest	Give possible reasons, explanations etc.
Show	Indicate or explain.
Analyse	Examine in detail in order to find the meaning/discover the essential features.
Why?/Explain/Account for/ Give reasons	Provide the causes of a feature, process or pattern. Explanation usually requires an understanding of processes. It is a higher-level skill than description and this is reflected in its greater mark weighting in examination questions.
To what extent?/Assess/ Evaluate/Discuss	These commands are evaluative. You need to consider the evidence connected to an issue/problem/technique, make reasoned judgements and present a viewpoint. This is the highest-level skill required at A2 and is widely used in extended-writing questions in F763 and F764.

Synoptic assessment

The Geographical Skills unit (F764), along with the Global Issues unit (F763), are the instruments used for synoptic assessment at A-level. The specification defines synoptic assessment as:

> ...assessment of candidates' ability to draw on their understanding of the connections between different aspects of geography represented in the specification, and to demonstrate their ability to 'think like a geographer'.

Synoptic assessment requires you to use knowledge, understanding and skills drawn from outside the content of an A2 unit — that is, from studies at AS and elsewhere at A2. Synopticity is an intrinsic part of Geographical Skills. Inevitably we draw on knowledge and understanding of geographical environments (e.g. rivers, coasts, cities) to apply investigative methods such as hypothesis formulation and data analysis. It is also likely that you have conducted fieldwork and research investigations at AS and for the A2 Global Issues unit. Drawing on this experience is clearly in line with the specification's definition of synopticity.

Synoptic assessment has two main purposes. First, it encourages you to adopt a broad perspective when analysing people and environment problems. This is a quintessential geographical approach, seeking to integrate (and synthesise) your understanding of society, economy and the physical environment. Second, it gives the A-level geography specification a coherence which, because of modularity and the reduction of the subject into a series of discrete units, it might otherwise lack.

About this section

The Questions and Answers section provides student answers to structured data-response and extended-writing questions. These questions cover the six stages of geographical investigation described in the **Content Guidance** section of this guide. There are eight exam-style structured questions and six extended-writing questions. Each student answer is assessed against the mark schemes on pages 35–36.

Examiner's comments

Examiner comments on the questions are preceded by the icon ⓔ. They offer tips on what you need to do in order to gain full marks. All candidate responses are followed by examiner's comments, indicated by the icon ⓔ, which show how marks have been awarded and highlight areas of credit and weakness. For weaker answers the comments suggest areas for improvement, by highlighting specific problems and common errors such as lack of development, excessive generalisation and irrelevance. Careful scrutiny of these comments is highly recommended. They will ensure that when you sit the exam you understand how your answers will be marked and exactly what the examiners are looking for.

Section A: Structured data-response questions

Question 1 Identifying a hypothesis and collecting primary and secondary data

Figure 1 The coastline near St Bee's Head, Cumbria

Study Figure 1, a photograph that shows a potential location for an A-level geography investigation.

(a) (i) State and justify, using evidence from the photograph, an appropriate question or hypothesis for geographical investigation at this location. (5 marks)

  (ii) Describe and justify how you would collect the primary data needed for the investigation you have chosen. (10 marks)

(b) Assess the value of two secondary data sources that might contribute to the investigation. (5 marks)

ⓔ Successful answers will (a) relate the evidence of the photograph to all three parts of the question, (b) provide details of data collection strategies and (c) ensure that command words such as 'justify' and 'assess' are given due prominence.

Student A

(a) (i) A possible question to investigate would be the effect the groynes have had on beaches **a**. Groynes are wooden fences that run at right angles to the coast and interrupt the process of longshore drift **b**. The photo shows a small groyne field extending along this stretch of coast **b**.

ⓔ **3/5 marks awarded**   **a** The answer suggests an appropriate question but fails to use the photograph to justify such an investigation at this location. **b** Most of the answer comprises description, which is not directly relevant to the question. Without justification for the proposed question, this is a Level 1 answer. Justification from photographic evidence might have included references to: (1) only part of the coastline being protected by groynes, offering scope for a comparative study; (2) the scale of the study area; (3) the presence of beaches; (4) accessibility of the site.

(ii) Groynes are designed to encourage the deposition of sand and shingle on coasts where longshore drift occurs. By extending out into the sea, at right angles to the coast, groynes intercept waves and currents. As the volume of beach sand and shingle increases it helps to protect the coastline from erosion. **a**

In order to investigate the question 'do groynes have an affect on beaches?' I would collect primary data by surveying the width and cross-sectional shape of beaches in the area protected by the groynes. There I would use two methods of data collection. Beach width would be measured with a tape, following the direction of the groynes. This would be done at low tide when the beaches had maximum width. Next I would measure the beach profiles. This would also be done at low tide. Measurements of beach profiles would be achieved by using a tape, ranging poles and a clinometer. Starting at the top of the beach, I would measure slopes angles every 10 metres (10 metre sections would be shown by ranging poles). The angle between each pole would then be measured using a clinometer. It is important to ensure that the poles are vertical and that slope measures are taken between the same positions on each pole. **b**

Having collected the data from beaches protected by the groynes, I would take similar measurements in the area shown on the photograph beyond the groyne field, where erosion of the cliffs is taking place. These beach data could then be compared with those collected from the beaches in the groyne field to see if there were any differences between the widths and profiles of the two types of beaches. **c**

ⓔ **5/10 marks awarded**   **a** The initial paragraph is peripheral to the focus of the question, which is primary data collection. A more effective strategy would deal directly, and from the outset, with data collection methods. **b** The answer provides no detail of sampling techniques such as the location of beach transects and the number of transects needed. **c** The answer fails to provide any justification for the choice of the data collection strategy described. Although two appropriate methods of data collection are outlined, the lack of sampling detail and any justification for the strategy used, suggests that this answer is Level 2. The response is, however, soundly structured, with few errors of grammar and spelling and some use of geographical terminology.

(b) Two possible secondary sources that might contribute to the investigation are maps and photographs. Old maps might show how the coastline has changed since the construction of the groynes **a**. So too might old photographs. Old aerial photographs **a** could be particularly useful because they might show differences in the width of beaches between the groyne field and the stretch of coast that is not defended by groynes **b,c**.

ⓔ **2/5 marks awarded** **a** There is a clear understanding of secondary sources and two valid secondary sources are identified. **b** Some attempt is made to show how secondary sources could be used but this could have been elaborated in a little more detail. **c** Only limited attempts are made to evaluate the usefulness of the stated secondary data sources. Details of specific map types are not mentioned (e.g. map scales, dates of publication) and there is only limited detail in the description of photographic sources. The absence of any significant evaluation and the limited amount of detail provided on the secondary sources indicate a Level 1 answer.

ⓔ Overall this answer scores 10/20 marks — a D grade.

Student B

(a) **(i)** An appropriate question for geographical investigation in the area shown by the photograph is: are beaches wider and steeper in the area protected by groynes, than elsewhere? **a** The justification for this question **b**, from the evidence of the photograph, includes: the contrast over a short distance between the protected and non-protected sections of coast; ease of access to the coast (this is a tourism area — e.g. caravan park, car parks, there are people on the beach, the sea wall is low) **c**. Because groynes are designed to intercept sediment transported by waves and currents, in theory the beaches protected by them should be steeper and wider than those in unprotected areas such as the far right of the photo **d**.

ⓔ **5/5 marks awarded** **a** At the outset the answer provides an appropriate question for investigation. **b** The key command word 'justify' is addressed immediately, demonstrating a focused and relevant approach. **c** The answer makes clear use of evidence on the photo relating to access to the shoreline and the contrasting sections of coast (a) protected (b) not protected by groynes. **d** The final sentence provides a rationale for the investigation. Though it is not essential, it shows sound understanding of the topic chosen for investigation. Overall, this is a clear and relevant answer, near the top of Level 2.

(ii) I would begin the investigation by defining two contrasting sections of coastline: first the section between the groynes; and second the section in the distance (on the far right of the photograph) which is not protected by groynes. In these two areas I would collect primary data relating to the width and the gradient on the beaches **a**.

Primary data would be collected by spatial sampling, using line transects. Sampling is needed because the area and length of the coastline are too large (given the limited resources and time available) for a complete survey **b**.

Also, if the sampling is done accurately, using objective scientific methods, it should produce enough data to answer the question in the investigation b.

Before collecting the primary data I would decide on a sampling strategy. The preferred sampling strategy is systematic c, based on transects c running parallel to the groynes, starting at the back of the beach (i.e. the sea wall) and ending at the low tide mark. Ten transects in each of the two sections of coastline would provide sufficient data to answer the research question a. At the coastline protected by groynes, transects will be located at the mid-point between each pair of groynes. In order to get a total sample of 10 transects, additional transects will be taken at the mid-point distance on the left side of the nearest, and the right side of the furthest groyne. The total length of shoreline between the 10 transects will then be measured, and this same length used for the coastline beyond the groyne field. There, transects will be defined at the same intervals to those used in the groyne field a.

The next step is to take measurements for each transect. This would be done at low tide when the beaches are fully exposed. Starting at the back of the beach, 10 metre intervals are measured with a tape and ranging poles. Ranging poles are placed between each 10 metre section and the slope angle between them measured and recorded using an instrument known as a clinometer c. If each transect is approximately 60 metres long, this would give 60 slope angles for the groyne-protected beach, and 60 for the beach not protected by groynes a. This amount of data should be enough to provide an accurate answer to the research question b. Measurement along the transect continues until the low water mark is reached. At this point the total length of the transect is recorded. The same method is then repeated for the other 9 transects in the groyne field, and later, the 10 transects outside the field.

Once data collection is complete there will be two sets of sample measurements (10 for beach width and around 60 for beach angles) for each of the two beach sections. This should be enough for data presentation and analysis to determine whether the two beaches are significantly different and whether groynes have an influence on beach size and shape.

e **9/10 marks awarded**   a The answer describes two aspects of appropriate data collection strategy accurately and in some detail. b There is clear justification for the strategies chosen which is well linked to the question under review. c There is good use of appropriate terminology. This is a Level 3 answer. It is clear, detailed and well structured, with accurate grammar and spelling. There is possibly a little more scope for justification (e.g. planning issues and timing) but overall this is a quality answer.

(b) Historic maps and Google Earth are two secondary data sources that might contribute to the investigation **a**. A mid-nineteenth century OS map of the coastline at a scale of approximately 1:10,000 could provide valuable background information on the coastline as well as coastal defences and rates of erosion. However, its value would depend partly on when the groynes were constructed. A map which shows the coast before the construction of the groynes would be most valuable **b**.

Google Earth provides a recent, large-scale satellite image of the coastline and could be valuable in assessing the extent of beaches, the amount of erosion and the influence of the groynes. It would also place this short stretch of coastline into a wider context and assist in understanding the marine processes operating there **b**. The value of Google Earth images will, however, depend on the state of the tide at the time of the satellite pass. If it is high tide, the images will provide little information on beach extent **b**.

ⓔ **5/5 marks awarded** **a** The answer focuses clearly on two possible sources of secondary data. It describes accurately these data sources and the information available and provides some relevant examples. **b** Some critical evaluation of each data source is attempted, with appropriate exemplification. This answer does exactly what is required, provides an appropriate level of detail, and merits a top Level 2 score.

ⓔ Overall, the answer scores 19/20, making it a very good A grade.

Question 2 **Data presentation and data analysis**

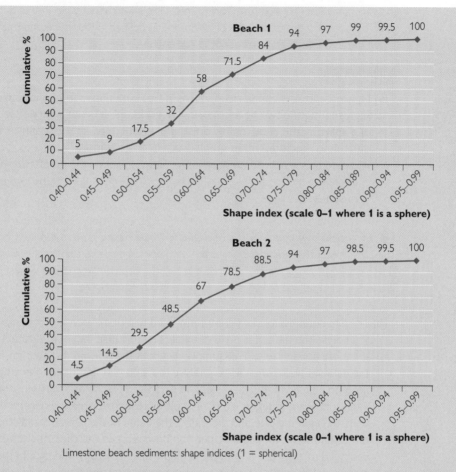

Limestone beach sediments: shape indices (1 = spherical)

Figure 2   Sediment shape on the Kent estuary, Cumbria

Figure 2 presents some of the results of a geographical investigation into the shape of limestone shingle on two beaches near the mouth of the Kent estuary in Cumbria, backed by low limestone cliffs. A sample of 200 particles was collected on each beach in order to answer the following question:

'Are there significant differences in the shape of limestone shingle between the two beaches?'

(a) Comment on the effectiveness of Figure 2 in representing the data used to answer the research question. (5 marks)

(b) Describe and explain one relevant statistical technique that could be used to analyse differences in particle shape between the two beaches. (10 marks)

(c) Suggest possible reasons why fieldwork and/or research investigations often fail to show the results you expected. (5 marks)

(e) Commenting on the effectiveness of a specific graphical or mapping technique (question (a)) is an evaluative exercise. You are judging a technique, and to do this you need to assess

effectiveness against a number of criteria. These criteria might include: visual impact; the extent to which patterns and trends are made clear; the degree of generalisation; and the ability to extract accurate numerical data from the chart or map. The key to a successful answer to question (b) is to recognise that statistical analyses of 'difference', as defined in the specification, refer to either the U test or the Chi-squared test, but not to correlation.

Student A

(a) Figure 2 shows a comparison between two cumulative frequency curves representing the two beaches being investigated. It is not a very effective method of displaying the data, mainly due to the visual impact. The graphs are more difficult to read than other graphs such as histograms **a**. Both look identical and it is hard to see any differences in the data. However, the graphs are useful in challenging the hypothesis and showing that the shingle on the two beaches is very similar **b**.

ⓔ **2/5 marks awarded**   **a** The reference to 'visual impact' is not elaborated, nor is the comparison with histograms explained. **b** A valid point is made concerning the similarity between the charts, suggesting a minimal difference in particle shape between the beaches. This is a Level 1 answer — it is brief and provides only limited comments on Figure 2.

(b) The Chi-squared test is a commonly used test of difference and can be used to analyse the data. It is a very flexible test because it is non-parametric and therefore doesn't rely on assumptions about the normal distribution from which the data set was drawn **a**. To analyse the data you first need to arrange the data for both beaches into classes **b**. The data used for the analysis must be absolute values, not percentages. The Chi-squared test compares the observed distribution with an expected one and the bigger the value of the Chi-squared statistic the more significant the result. Once you have worked out the Chi-squared value **c** you need to compare it with the critical value at 95% confidence in the appropriate statistical tables. If the value is greater than the critical value it is statistically significant. This would mean that there is a real difference in the shape of particles between the two beaches.

ⓔ **7/10 marks awarded**   **a** The answer makes important points about the relevance of the Chi-squared test to the data. **b** There is a lack of detail in areas such as number of classes and the number of values in each class. **c** Little insight is provided on the calculation of Chi-squared. The answer is relevant, has sound detail and accuracy and is soundly structured with few errors of grammar and spelling. There is some valid use of geographical terminology. However, the answer is rather brief and ignores some important areas.

(c) One of the main reasons why fieldwork often fails to show what you were expecting is due to human error **a**. This is particularly apparent in investigations where you are collecting data with other people. This can often cause problems as different people may record data to different degrees of accuracy or observe things differently. Information such as sediment roundness is especially vulnerable to human error as it is an entirely subjective measurement. This has happened in some of my past fieldwork where there has been little visual difference between the categories of roundness of particles **b**.

Another reason why results might not fit your hypothesis could be because not enough planning has gone into the investigation **a**. If you don't develop a solid theory for your hypothesis it is likely that your results will not be what you expected. It is possible that even when you have planned an investigation thoroughly there may be factors that cannot be accounted for **a**. For example, if you intended to study the bedload of a river and there was a storm a few days before you collected your data, your results could be distorted. There may also be unforeseen human influences in the area such as the dredging of river channels **b**. Additionally you might arrive at your fieldwork site and find there isn't as much information available as you expected e.g. if there aren't enough people to interview in a town centre. **c**

ⓔ **3/5 marks awarded**   **a** The answer makes three valid points but fails to develop them convincingly. **b** Exemplification is often weak. **c** The answer is rather long — longer than (b), which is worth twice as many marks! A more effective structure would involve stating two or three factors at the outset (i.e. human error, lack of data, unexpected change) and then briefly developing each factor, supported by an example. The answer has not quite got the balance right and fails to take full account of the mark weighting for each question. This is a solid Level I response.

ⓔ The total score for this question is 12/20 — a grade C.

Student B

(a) At first glance the two cumulative frequency curves are of similar shape, but on closer examination differences can be seen. Of the 200 particles on each beach those on beach 2 tend to be more spherical **a**. This explains the more constant and steady rise of the cumulative frequency curve for beach 2, whereas the curve for beach 1 has a sharp increase in gradient between 0.55 and 0.64 **b**. Although I would not class the differences in the shape of limestone shingle as significant, differences do exist **b**. The charts are in a sense effective because they highlight these differences, even though they appear to be quite small **a**. On the other hand, because the charts are very similar in shape, an alternative method such as a simple frequency curve, which included values for the mean and standard deviation, might show the differences even more clearly.

ⓔ **4/5 marks awarded**   **a** Only two sentences deal explicitly with the effectiveness of the charts. **b** The answer shows careful analysis and interpretation of the two charts. Apart from the final sentence, this is a clear and relevant answer. However, a little more balance, with more references to Figure 2, is needed to achieve a top mark.

(b) In order to establish statistically significant differences in particle shape between the two beaches a statistical test such as Chi-squared could be applied to the data. In this way the two frequency distributions can be compared statistically **a**. Data must be converted from percentages back to absolute values for the analysis **a**. The greater the value of the Chi-squared statistic the more that the differences between the two distributions are not due to chance. The significance of Chi-squared is determined by comparing its value with the critical value at the 95% confidence level, in Chi-squared tables. If the Chi-squared value exceeds the critical value at 95% we can say that it is statistically significant.

As an alternative, we could calculate the mean, median and modal values and compare the two data sets. We could also calculate the standard deviation which would show the clustering or dispersion of values around the mean. The higher the standard deviation the more spread out the data are **b**.

ⓔ **6/10 marks awarded** **a** The description of Chi-squared is accurate and appropriate, but generalised and somewhat brief. There are no details of calculation; nor is there comment on the non-parametric nature of the test and the importance of degrees of freedom. **b** As the question asks for one relevant statistical technique only, the second paragraph is irrelevant. Marks are awarded for the description of Chi-squared and for the answer's sound structure and relatively few errors of grammar and spelling. However, the answer's limited detail restricts it to a modest Level 2.

(c) One of the main limitations of fieldwork/research investigations is methodological **a**, both in the way an investigation is planned and how it is carried out in the field. Investigations may not produce the results you expect as your expectations may be highly idealised and due to the complex nature of the real world **a**, outcomes are hard, if not impossible, to predict. Challenges such as obtaining a representative sample of a population **a** are difficult as an adequate sample size can be limited by research time constraints. An example would be the difficulty of achieving representative samples of shoppers (i.e. age, gender, income) in street interviews where rejection rates are high, and where footfall is relatively low **b**. Often factors that limit the accuracy of an investigation are overlooked. For example, inadequate equipment **a** — the use of a float rather than a current meter in measurements of stream velocity **b**.

ⓔ **4/5 marks awarded** **a** The answer suggests a number of valid reasons why the results of fieldwork/research investigations often fail to produce expected results. **b** These reasons are supported by a couple of plausible examples. As a whole the answer is fairly detailed and accurate and just about satisfies the Level 2 criteria.

ⓔ The total score for this question is 14/20 — a grade B.

Question 3 Data presentation, statistical significance and sampling

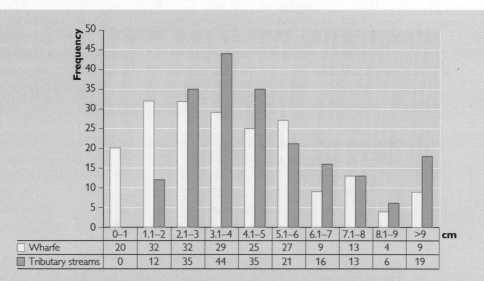

	0–1	1.1–2	2.1–3	3.1–4	4.1–5	5.1–6	6.1–7	7.1–8	8.1–9	>9	cm
Wharfe	20	32	32	29	25	27	9	13	4	9	
Tributary streams	0	12	35	44	35	21	16	13	6	19	

Figure 3 Sediment size on the River Wharfe and tributaries

Figure 3 presents some of the results of geographical research into the size of river sediments. The data, collected by systematic sampling in the field, aim to answer the following question:

'Are the sandstone bedload sediments in the River Wharfe smaller than those input by shorter tributary streams?'

(a) Comment on the effectiveness of Figure 3 in representing the data used to answer the research question. (5 marks)

(b) Explain why it is often important to establish the statistical significance of the differences between two data sets and describe how you would do this. (10 marks)

(c) Why is it often difficult to obtain accurate and representative samples in geographical investigations? (5 marks)

ⓔ Statistical significance is a concept that lies at heart of numerical data analysis in geography. It is frequently misunderstood and often ignored by students. In question (b) a sound understanding of the relationship between statistical significance and sampling methodology is essential for a successful answer.

Student A

(a) Figure 3 is a type of histogram **a**. If it is effective it should allow us to pick out the patterns and differences and help us decide whether the bedload particles in the River Wharfe are smaller than those in the tributary streams **a**. We can immediately see that the larger particles are found in the tributary streams and the smaller ones are in the River Wharfe **a**. Comparison is aided because the bars representing the tributaries and the River Wharfe are side-by-side in the same categories. The histogram also allows us to compare the size frequencies of the two distributions.

🅔 **3/5 marks awarded** **a** The answer makes a number of valid points: it correctly identifies the chart as a histogram; argues that the data are clearly presented and are therefore easy to interpret; and sees the scope of the chart for comparative purposes. But overall, the answer has insufficient focus on the advantages and limitations of the chart to justify Level 2. For instance, there is little direct reference to: the data presented in Figure 3; problems of data generalisation in classes; and the fact that the one class is open-ended. This is a limited answer, which achieves Level 1.

(b) Establishing a significant difference between two sets of data allows us to say that the difference has not occurred by chance. It is essential to know this when writing the analysis or conclusion to an investigation **a**. If there is no statistically significant difference between the data sets then our hypothesis is rejected in favour of the null hypothesis (i.e. hypothesis of no difference). But if the difference is established as significant then we have confidence in accepting the hypothesis. The reason why we have to test for significance is because investigations like the one in Figure 3 rely on sample data and we need to know that any differences have not occurred purely by chance **a**.

In this particular investigation Chi-squared **b** would be a suitable test of difference as the data are grouped in frequencies **c**. First the null hypothesis would be stated which in this case would be: 'there is no significant difference in the size of sandstone particles between the River Wharfe and its tributaries'. As there are 12 classes in the distribution (some with very few values) it would be appropriate to reduce these to just 6 or 7, and aggregate some of the data. This is because the Chi-squared test becomes unreliable if several classes have expected values of less than 5 **c**. The Chi-squared value is then obtained by finding the expected frequencies (assuming the values were distributed randomly) and substituting them in the formula:

$$\text{Chi-squared} = \sum \frac{(\text{Observed–Expected})^2}{\text{Expected } \textbf{c}}$$

The significance of the Chi-squared value is then obtained from statistical tables **c**. If the critical value at the 95% confidence level is less than the Chi-squared value, the result is said to be statistically significant.

🅔 **9/10 marks awarded** **a** The answer demonstrates a clear understanding of the concept of statistical significance. **b** The answer identifies a relevant statistical technique. **c** Chi-squared is described clearly and accurately and a detailed explanation of its value in this context is given.

This is a good Level 3 answer. It is well structured with accurate use of grammar and spelling and makes good use of geographical terminology. For maximum marks a fuller explanation of how expected frequency values are obtained, and the importance of degrees of freedom in determining statistical significance, could be added.

> **(c)** A sample is a small proportion of a population that provides the data for an investigation. A weak sampling strategy is likely to result in inaccurate data which does not represent the population **a**. This will undermine the validity of the investigation. Human error in data collection (e.g. measurements, interviews) can also lead to inaccurate data **b**. There are often many uncontrollable variables in geographical investigations that adversely affect sampling **b**. For example, when collecting river sediments, bedload particles may have been dislodged or moved by people. There is also the temptation to select particular particles because they are easy to obtain and measure, rather than adhere to a strict sampling method.

ⓔ **2/5 marks awarded**   **a** In a short-answer question it is important to get straight to the point. The first two sentences fail to do this and are marginal to the question. **b** The answer refers to some reasons why samples are unrepresentative (e.g. human error, bias) but fails to develop them fully. The answer would be credible if it were based around specific examples. For instance, the problems of interviewing, the fact that some populations in physical geography are not distributed randomly, and issues of sample stratification. This is a Level 1 answer.

ⓔ The total score for this question is 14/20 — a grade B.

> **Student B**

> **(a)** Figure 3 would seem to represent the data fairly well. We can see the clear comparison between the size distribution of particles in the River Wharfe and its tributary streams and anomalies such as the data in class 1.1–2 **a**. However, the chart does not tell us where the samples were collected and this makes comparison of the distributions difficult **b**. Nor does it give us information on individual tributaries **b**. The hypothesis clearly makes reference to the length of streams being a factor influencing differences between the two distributions, but the chart makes it difficult for us to judge the reason for the results **b**. Overall, drawing any conclusions from the chart is problematic.

ⓔ **1/5 marks awarded**   **a** Only the first two sentences are relevant to the question. **b** Most of the answer fails to address the question. Instead of focusing on the suitability of Figure 3 as a technique for presenting the data, it gets sidetracked into issues of data collection strategies. It also misinterprets the research question. In general it is a struggle to find anything to reward in this limited response.

> **(b)** It is often very important to establish the statistical significance of the difference between two sets of data. This could be because the data sets were collected under different conditions. For example beach profiles might be measured at different stages of the tide or different sampling methods might have been used to collect the data (e.g. random, systematic). Human errors in measurement, transcription and recording may also occur. **a**

Establishing the statistical significance allows us to assess how reliable the difference between the two data sets is and will allow us to prove or disprove a hypothesis, or answer a question **b**. It also tells us how much confidence we can have in the results and that if we were to repeat them, whether we would get similar results. **b**

One of the tests that can be used to find the significance of the difference between two sets of data is Chi-squared. The data showing the two distributions would be analysed using Chi-squared. The Chi-squared statistic would then be looked up in tables alongside degrees of freedom (calculated by multiplying the number of columns minus one by the number of rows minus one) at the 95% confidence level. **c**

e 4/10 marks awarded **a** The first paragraph contains little of relevance to the question. **b** Some understanding of the concept of statistical significance is evident in the second paragraph and is credited. However, development (e.g. through exemplification) is lacking. **c** An appropriate statistical test — Chi-squared — is stated, but only a limited attempt is made to show how Chi-squared is used. Although the answer is soundly structured, with relatively few errors of grammar and spelling, its overall brevity, together with irrelevance, lack of exemplification and limited development, indicate a modest Level 1 standard.

(c) There are many reasons why it is often difficult to obtain accurate and representative samples in geographical investigations. Two important reasons are unexpected physical events and human error. A naturally occurring event such as a river flood could affect the accuracy and representativeness of sediment samples, especially if samples were taken over a period of several days. Floods will tend to remove finer sediment and deposit coarse sediments over a wide area of the river channel. In the uplands, floods may cause erosion and mass movement which input an unusual amount of coarse sediment into a river. **a**

Human error in measurement and recording can also reduce the accuracy of samples. For instance, measuring the slope of rivers with a shallow gradient with a clinometer can produce widely varying results. Estimating subjectively environmental quality in a town or city can be equally inaccurate. **b**

e 3/5 marks awarded **a** The answer outlines two valid reasons and the first paragraph provides some detailed and relevant exemplification. However, the example, and its relevance to the question, are not fully explained. **b** Exemplification of human error in sampling is rather vague and, as in the first paragraph, the example needs more explicit linking to the question. The answer does not address the problem of representative sampling. At best this is a partial answer, which overlooks obvious difficulties such as the timing of interviews, the cooperation and honesty of respondents, and the multivariate nature of geographical problems, resources, scale and so on. These weaknesses make it a Level 1 rather than a Level 2 response.

e The total score for this question is 8/20 — a grade E.

Question 4 **Statistical mapping methods**

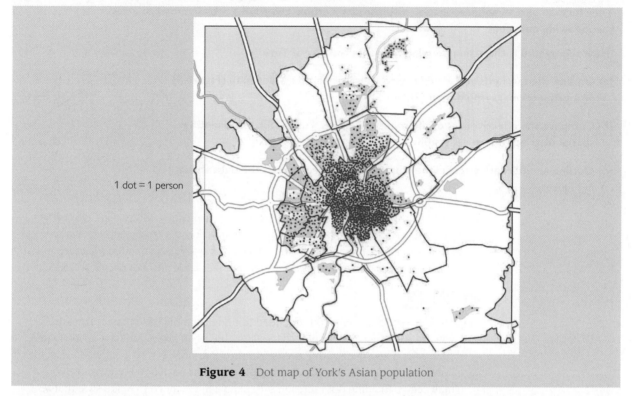

1 dot = 1 person

Figure 4   Dot map of York's Asian population

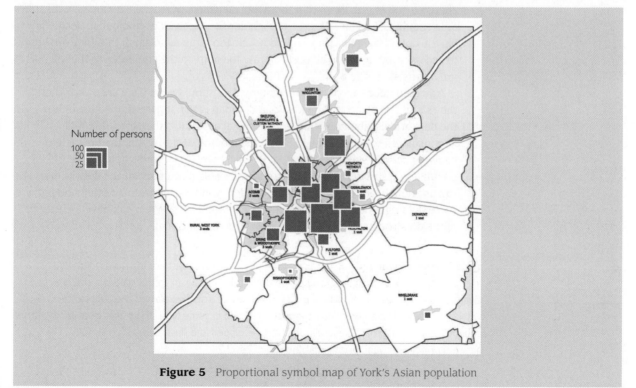

Number of persons

100
50
25

Figure 5   Proportional symbol map of York's Asian population

Figures 4 and 5 show the distribution of the Asian population in York, by ward, in 2001. They formed part of an investigation into the city's ethnic geography, which aimed to answer the following question:

'How segregated is the Asian population within the city of York?'

(a) Outline the methods you would use to construct either a dot map (Figure 4) or a proportional symbol map (Figure 5). (5 marks)

(b) Compare the effectiveness of the maps in Figures 4 and 5 for representing spatial distributions in geography. (10 marks)

(c) Outline an alternative technique you could use to represent the data shown in Figures 4 and 5. (5 marks)

ⓔ Any revision programme of data presentation techniques (i.e. statistical maps and charts) should cover the following: method of map/chart construction; the appropriateness of a particular technique in relation to its purpose and the nature of the available data; and an evaluation of a technique's strengths and weaknesses compared with alternative methods of data presentation.

Student A

(a) To construct a dot map you first need to find a suitable base map of the area you are studying. This map needs to show the boundaries you have investigated, such as census super output areas, wards or parishes a. Next you need to decide how many units of information are represented by each dot. For example you may decide that one dot represents 10 people b. You need to make sure that you choose a suitable scale so that there are a few dots in low density areas, while the dots should just begin to merge in high density areas. You should also decide how big the dots should be. They should not be too large or so small that they are hard to see or count. It would be best to draw the map on the computer so that all dots are a consistent size b.

You then need to work out how many dots should go in each area and where they should be placed c. The idea of a dot map is that it gives some idea of how the data are distributed within geographical areas. It is therefore essential to consult other types of map (e.g. an OS map) to decide where the dots (e.g. showing people) should be located. For instance, if you were showing the population distribution in a rural area you would place most of the dots in market towns and villages c. When placing the dots you should make sure that you put some dots right up to the edge of the area boundaries. When you have located all of the dots you need to add a title, key, direction and scale.

ⓔ **5/5 marks awarded** a The first step — finding a suitable base map — is identified and exemplified. b Defining the value of the dots and choosing an appropriate dot size are clearly stated and again exemplified. c The placement of dots is explained and is illustrated with an example. The answer describes accurately and in some detail the stages involved in the construction of a dot map. General points are developed and illustrated with examples. The answer is wholly relevant, placing it at the top of Level 2.

(b) Both dot maps and proportional symbol maps have advantages and disadvantages for displaying information. On the whole, for reasons of accuracy and visual impact, dot maps are more effective at showing population distributions.

The main advantage of dot maps is that they show the distribution of data within areal units such as wards and super output areas **a**. This gives you a much more accurate impression of the area you are studying. It also helps because it shows the transition between areas **b**. Many maps, such as the proportional symbol map, only show one generalised unit of data for an entire area, which may not be very helpful if the area is large **b**. Another advantage of dot maps is that they have a large visual impact **b**. The high density clusters of dots in the centre of York attract attention because they make it clear that the area has a high proportion of Asian people.

Dot maps are also very easy to use because all you need to do to interpret the map is count up the dots (although this may be difficult in high density areas where the dots begin to merge) **b**. Despite these advantages, dot maps have some disadvantages, chiefly in the length of time they take to construct. Placing all the dots accurately is time consuming, not only in drawing the map but also in researching where to put the dots in the first place **b**. Related to this is the problem that the dots, although placed in roughly the right area, are placed randomly and so do not represent the exact location of whatever they represent **a**.

The main advantage of proportional symbol maps is their immediate visual impact and the ease with which it is possible to identify large and small squares **b**. For example, in the centre of York the large squares make it quite clear that that a large Asian population is concentrated there. Conversely, it is clear that there is only a small Asian population around the southern edges of the city. However, these advantages are overshadowed by the disadvantages of proportional symbol maps.

The main disadvantage is that it is easy to underestimate the values that the symbols represent **b**. This is because doubling the sides of a square increases its area four times **a**. Another disadvantage is that proportional symbol maps don't show the internal distribution of the Asian population within the wards **b**. Finally, in areas of York with small wards such as the city centre, it becomes difficult to identify symbols and measure the length of their sides accurately.

🅮 **9/10 marks awarded**   **a** The answer demonstrates a clear and detailed understanding of dot and proportional symbol maps. **b** There is clear and accurate analysis of the advantages and disadvantages of the maps. This critique is effectively supported by examples and specific references to Figures 4 and 5. This is a substantial and very good Level 3 answer. It is also well structured with accurate grammar and spelling and there is good use of appropriate geographical terminology. However, a point-by-point comparison would have been preferable and would have placed more emphasis on critical comparison.

(c) A possible alternative mapping technique that could be used to show the data in Figures 4 and 5 is choropleth mapping. Choropleth maps show differences in values between areas by differences in shading (or colour) **a**. They are fairly easy to draw because data are usually available for the areal units on the map **b**. To construct a choropleth map you first need to have a base map which shows the areal units (e.g. wards, super output areas) **a**. You then need to divide the population totals into a number of categories and decide on the boundaries between them. In this example, with less than 20 areal units, four or five categories could be chosen **a**. Fewer than four would make the map too generalised; more than five could obscure important patterns and trends **b**. The final decision involves shading or colouring the map **a**. A black and white map would use darker tones to represent the higher values, and lighter tones to show the lower values. If the map were coloured, softer pastel colours (e.g. yellow and green) would show lower values and more striking colours (e.g. red and purple) the higher values **b**.

ⓔ **4/5 marks awarded** **a** The answer selects and outlines an appropriate alternative mapping technique. The description is clear, accurate and relevant. **b** The answer is effectively developed and exemplified in places. This is a solid Level 2 answer that does exactly what the question demands and is not drawn into unnecessary assessment or evaluation. A little more reference to the map's precise content (i.e. the Asian population of York) would give maximum marks.

ⓔ The total score for this question is 18/20 — a grade A.

Student B

(a) Proportional symbols use circles, squares, triangles and bars for mapping. They are based on the principle that the area of each symbol is proportional to the value it represents **a**. Unlike choropleth maps, proportional symbol maps show absolute values such as population counts and retail floorspace.

When drawing a proportional symbol map the first task is to choose a type of symbol. In most instances, either squares or circles are used **b**. The next and most important step is to choose the scale **b**. The size of the symbols should represent the data as accurately as possible so that anyone viewing the map can immediately gauge from the size of the symbols the values they show (see Figure 5) **c**. A carefully chosen scale will also help to ensure that symbols do not overlap excessively and obscure one another **c**.

Calculating the size of the symbols involves finding the square root of the values to determine the radii of circles or the sides of squares **b**. The symbols are then plotted centrally within each areal unit. Finally a key, title, scale and direction sign should be added to the map to make the use of the map intelligible to the reader.

ⓔ **4/5 marks awarded** **a** The answer begins with an accurate definition of proportional symbol maps and how they represent quantitative data. **b** This shows clear and accurate understanding of constructional methods. These are dealt with systematically, and the various steps are easily followed. **c** Some of the steps are expanded and developed effectively in more detail.

Overall, this is a good answer. Further improvement might involve some reference to base maps and to Figure 5, as well as detail on the calculation of the size of symbols.

(b) The dot map which represents the Asian population in York is effective in several ways. Firstly the data are shown individually, rather than being generalised in one symbol that covers a whole area a. This means that it is possible to get an accurate picture of the Asian population distribution within York. However, the problem with the dot map is that because there are large concentrations of population in some areas the map b can become overcrowded as the dots begin to merge a. This makes it difficult to recover the data and estimate the population in some places such as the central area a.

   The proportional symbol map has a very large range of values for each size category (i.e. 50). This means, for example, that there is no distinction in terms of symbol size for a value of 51 people or 99 people a,b. This makes the map fairly generalised and therefore reduces its accuracy compared with a dot map a. A further problem with the proportional symbol map is that in places the symbols overlap each other, making precise interpretation difficult a. Nonetheless, the symbols do give a fairly good idea of the overall spread of Asian people in the city a,b. In summary both maps have their plus and minus points.

(e) 7/10 marks awarded   a The answer provides a number of points relating to the strengths and weaknesses of dot and proportional symbol maps and draws some valid conclusions. b Credit is given for attempts to use Figures 4 and 5 as specific illustrations of general points. The answer could be improved by a more comparative evaluation of the two map types. The criteria used for comparison are not defined and references to the two maps of York could be extended. Overall the answer is sound and does a reasonable job. It is well structured with accurate grammar and spelling and some use of appropriate geographical terminology. However, it lacks the necessary detail to achieve Level 3. This is a top Level 2 answer.

(c) An alternative mapping technique to show the distribution of York's Asian population is isopleth mapping. This is where values of population density are shown by lines joining places of equal value. Usually this type of map is used in physical geography to show relief and temperature a. To draw an isopleth map you must first plot the values for population density as points on the map. Then it is necessary to decide on how many isopleths are needed b. Too many and the map will be difficult to draw and may lack clarity b. Too few and the map will be overgeneralised. In this example four or five isopleths would be sufficient. Inserting the isopleths between the point values is done by interpolation b. This is done on the assumption of a uniform change in values over distance between two points. For example if two point values of 80 persons/km^2 and 100 persons/km^2 are separated by 2 cm on the map, the 90 isopleth would be drawn exactly midway between the two points (i.e. at 1 cm from each). Care should be taken to ensure that isopleths don't touch or overlap b. They should also flow smoothly, without abrupt changes in direction. The map would be completed with a title, key, scale and direction.

ⓔ 5/5 marks awarded **a** The answer shows clear appreciation of the features, value and function of isopleth maps. **b** The main steps for constructing isopleth maps are well understood and clearly stated. This description is complemented by an awareness of some of the difficulties of isopleth mapping. This is a clear and detailed answer, of appropriate length, which does precisely what the question asks. Overall this is a top Level 2 answer.

ⓔ The total score for this question is 16/20 — a grade A.

Section B: Extended-writing questions

Question 5 Sampling methods in geographical investigation

Using your fieldwork and research experience at A-level, describe the use you made of sampling methods in data collection and assess their value.

(20 marks)

ⓔ It is essential to structure answers to questions in Section B around personal fieldwork/research investigations undertaken by students. Generalised answers, with infrequent references to personal investigations, must be avoided at all costs.

Student A

In a fieldwork investigation into the distribution of scree particles by size on a talus slope at Conistone in North Yorkshire, we made use of both random and systematic spatial sampling methods **a**. The aim of the study was to answer the question: 'do scree particles on a talus slope get bigger with increasing distance downslope?'

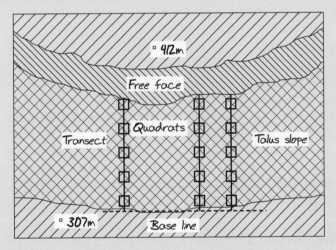

Spatial sampling on a talus slope

We started by establishing a base line of 50m at the foot of the slope (see sketch). Then, using the random number generator on our calculators we selected three locations at random intervals along the base line. Then we used tapes to delimit lines of transect perpendicular to the base line to the top of the slope. We then chose five sampling stations at regular intervals along each transect. The first station was at the base of the transect and the last one was at the top **b**. The stations were the sites for sampling scree particles. Equal intervals for the stations were chosen in order to get complete coverage of the slope and the stations were located on each transect at comparable heights. In this way it was hoped to get a representative sample of particles in relation to their position on the slope **b**.

The random factor used when selecting the position of the transects on the base line removed bias on behalf of the data collector, and was an attempt to use a fair and scientific approach. The selection of only three transects was dictated by limitations of time and the small number of students involved in the enquiry. Our confidence level in our final analysis could have been increased if we had taken a larger sample (e.g. four or five transects) **b**.

Stations for data collection on the slope were chosen at regular intervals. This systematic sample was used to gain a representative cross-section of particles. A random location for the sample stations was inappropriate because the site of each station on the slope was crucial to the investigation **b**. The final sampling method we used which actually led to the collection of data was quadrat sampling. At each station a 0.5 m square quadrat was placed on the slope and the scree particles selected for measurement beneath the quadrats' cross-wires. 50 particles were selected at each station and their median axes measured in centimetres with a ruler **b**.

These data were recorded at every station on the slope, and provided a large and objective sample on which to base our investigation. The process of data gathering using quadrats was also systematic. Instead of choosing particles subjectively, the particles were selected at regular intervals using the quadrat's cross-wires **b**.

The purpose of data collection was to gain an objective and representative sample of scree particles from a talus slope. To achieve this we used spatial sampling methods which involved both random and systematic elements and relied on transects and quadrats. Overall we collected 150 particles for each of five stations on the slope. This was a large enough sample to test the hypothesis and I was confident in the quality and reliability of the data **c**.

However, there were some problems which may have introduced small errors into the measurement of the scree particles. Some particles were too large to measure accurately and in places it was obvious that the scree had been disturbed by people and sheep. At the base of the slope some larger particles had been cleared to make a path, while particles that were partly buried beneath other particles were impossible to measure. The fact that our party was on the slope (which was unstable) also caused disturbance and sometimes it was difficult to anchor the quadrat in position (or remove particles without shifting the location of the quadrat). Despite these problems, the spatial sampling method was a success and achieved its purpose **c**.

ⓔ **19/20 marks awarded**   **a** This sentence clearly states to the examiner the fieldwork context of this investigation. **b** Descriptions and explanations of sampling methods are clear and detailed. **c** The last two paragraphs provide some detailed evaluation of sampling methods. This is a grade-A standard, Level 3 answer, which provides clear and detailed description, explanation and evaluation. The answer is firmly set within the context of specific fieldwork investigation and is well structured with accurate use of spelling, grammar and geographical terminology.

Student B

One piece of fieldwork that I carried out was a vegetation survey on a limestone ridge close to Morecambe Bay in Cumbria. The aim of the investigation was to determine whether differences in vegetation cover, vegetation height and vegetation types on a slope were influenced by small-scale changes in relief. Changes in relief affected exposure to the wind, soil depth and the availability of moisture a.

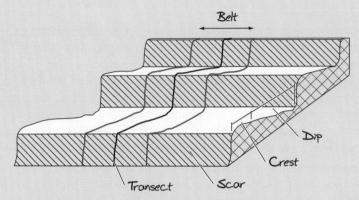

Spatial sampling of vegetation on Whitbarrow

To start with, in the study area, we defined three types of slope unit (see sketch): a dip slope, a slope crest, and a scar. Having done this we then laid out a 50 m base line at the top of the ridge and using random numbers between 1 and 50, selected the location of my transect, at right angles to the ridge. The transect was approximately 100 m in length b.

Because the line of transect was divided into a succession of dip, crest and scar units, samples were taken from each. These were done at systematic intervals at the mid-point of each slope unit using a quadrat b.

For each sampling point we counted the number of different plant species in each 0.5 m square quadrat and entered the results in a table like Table 1.

Table 1 Slope units and number of plant species

	Dip	Crest	Scar
5+			
3–4			
1–2			
0			

On average there were roughly three species per quadrat. They included bracken, blue moor grass, ling, moss and several other species b. This process was not as accurate as I would have liked. Some plants like bracken were large and a single plant could cover half or more of the squares. Also in places, the vegetation was in layers and some species like moss could only be seen by moving the surface layer c. We also recorded the vegetation cover in each quadrat and entered these data in a table like Table 2.

Table 2 Slope units and vegetation cover

	Dip	Crest	Scar
Complete cover			
Partial cover			
No cover			

This was done by counting the number of 10 cm × 10 cm squares within each quadrat where there was complete plant cover, partial cover and no plant cover **b**. This measurement was also open to error and often very subjective. Finally, judging the extent of the three slope units by eye in the field proved difficult. While the scar was very clear, it wasn't always obvious where the dip slope ended and the slope crest began. All of these problems were compounded because other groups, collecting data along their own transects, were making similar subjective judgements. As data from all the groups were used for analysis some error in the sample data was inevitable **c**.

Data on vegetation height were collected along the line of transect, but this time a belt transect was used. A zone 10 m wide on either side of the transect and parallel to it was defined. Within this belt transect we measured and recorded the maximum height of all juniper shrubs. They ranged in height from nearly three metres in the most sheltered spots to less than 35 cm in the most exposed places on the scars and slope crests **b**. Measuring the heights of the taller bushes was difficult as we only had metre rulers and ranging poles were only marked at 50 cm intervals. Where the vegetation was dense, access was also difficult. This was made worse by the prickly leaves of the juniper **c**.

In general, the sampling method, which was a systematic spatial sample using transects, belts and quadrats, proved fairly successful. Data were collected allowing us to analyse the affect of relief on vegetation. The main weakness was the subjectivity of some of the measurements/counts and this was made worse because data collected by other groups were shared to provide a large enough sample for statistical analysis. The problem of subjectivity and human error could have been reduced if each group had completed three or four transects and amassed their own data bases. However, there was insufficient time available to do this **c**.

ⓔ **19/20 marks awarded** . **a** The first paragraph clearly describes the location and purpose of the fieldwork investigation. **b** Descriptions and explanations of sampling strategies are clear and detailed. **c** Evaluations of sampling strategies are interspersed with description and explanation, and are equally clear and detailed. This is a grade-A standard, Level 3 answer. It is detailed and constructed around an original piece of fieldwork. Knowledge and understanding of sampling methods are clear, and there is a critical appreciation of some of the shortcomings of the techniques used. A sketch diagram and the inclusion of data recording tables help explain the purpose of the investigation and the methodology. The answer is well structured, with accurate grammar and spelling and good use of appropriate geographical terminology.

Question 6 Analytical techniques in geographical investigation

> State the nature of an investigation (fieldwork/research) conducted during your A-level studies. Describe the analytical techniques used in your investigation, justify your choice, and assess their usefulness.
>
> (20 marks)

ⓔ Extended-writing questions based around personal investigations invariably require some evaluation of methodologies. All too often, the importance of evaluation is neglected, and disproportionate emphasis given to description and explanation. Although evaluation is one of four tasks demanded by this question, a successful answer will give it some prominence.

Student A

During my A-level studies I undertook a fieldwork investigation into the spatial distribution of shoppers' spending in the CBD of Preston, in Lancashire a. One analytical technique I used was to calculate the total spending of shoppers in each of the six main shopping areas/streets in the city centre. From this I worked out the percentage share of total expenditure for each area/street and presented the information as a pie chart b.

I also calculated the mean of how much shoppers had spent in each area/street. This was useful in summarising each of the six data sets. In addition I calculated the standard deviation for each data set, allowing me to see the variation in shopper spending around the mean. This was important because it showed me how representative the mean was of each data set b.

In order to investigate further the spatial distribution of shoppers' spending I analysed data on the number of shoppers who made purchases in each of the six shopping areas/streets with the Chi-squared test. I did this because although I could see that the distribution of spending varied considerably between areas/streets in the city centre, I could not say with confidence that this result had not arisen by chance c. As this distribution was key to the investigation, establishing its statistical significance was essential c. To calculate Chi-squared you must first calculate the expected values. None of the expected values should equal zero, and no more than 20% should be less than 5. Chi-squared is calculated by subtracting each expected value from its matching observed value, squaring the result, and dividing by the expected value. You then sum all of the calculations (in this case six — one for each shopping area/ street). Finally you refer the Chi-squared values to statistical tables. If it is larger than the critical value it is statistically significant. Most of the time in geography we use the 95% level as being statistically significant c. In this investigation my Chi-squared value exceeded the critical value in the tables and enabled me to accept my hypothesis.

Another analytical technique I used in the investigation was percentages. In each shopping area/street I counted the number of comparison and convenience stores. I then calculated the percentages of both these shop types in each shopping area/ street. This was useful because it allowed me to see whether the shop type in each shopping area affected the amount of money that shoppers spent d.

In conclusion the most useful statistical technique I used was Chi-squared as the result of this test ultimately decided whether I rejected or accepted my hypothesis. However, without calculating descriptive statistics such as the mean, I would not have been able to analyse fully the data collected **d**.

ⓔ 12/20 marks awarded **a** The location and aim of the study are stated at the outset. **b** Largely a statement of some of the analytical techniques used. There is minimal description and explanation. **c** The Chi-squared test is clearly described and explained in some detail. **d** Limited attempts at evaluation. The answer has some clear and accurate description and evaluation of a number of relevant analytical techniques. They are linked clearly to an actual fieldwork investigation. Some assessment of the value of the techniques is given, though the emphasis is on description. The evaluation of these analytical techniques is very limited. Details of the investigation, with specific reference to places and the area studied, are also sparse. The answer is soundly structured with few errors of spelling and grammar. Use of geographical terminology is generally good. Despite its merits, the failure to provide clear and detailed evaluation makes this is a grade-C standard, Level 2 answer.

Student B

My enquiry looked at the effect of groynes on beach profiles at Overstrand in Norfolk. Groynes are used to stop coastal erosion and are an important feature of management along this stretch of the Norfolk coast. The hypothesis tested was that 'there is a significant difference in the profile of beaches protected by groynes, and adjacent unprotected beaches'. The reasoning is that groynes intercept sand and shingle carried by longshore currents and encourage deposition. This should result in a beach profile that is steeper, higher and longer than on an unprotected beach. At Overstrand both beaches consisted of a mixture of sand and shingle. **a**

Data were collected by surveying beach profiles using clinometers, ranging poles and tapes. We recorded slope angles along two transects with a clinometer at 5-metre intervals — first along a protected beach and then along an unprotected beach. The protected beach had 20 5-metre units and the unprotected beach had 12. **b**

We used a number of techniques to analyse the data in this investigation. We plotted the beach profiles onto graph paper. This allowed us to compare their gradient, length and height **c**. These profiles gave a good visual comparison and provided a starting point for further analysis **d**. They showed that although the protected beach was much longer than the unprotected beach (109 m and 67 m respectively) the average gradient appeared to be steeper on the unprotected beach.

We also drew the individual 5-metre gradients for each profile as dispersion diagrams. The single axis of dispersion diagrams allowed further comparison of the two data sets and again suggested that the unprotected beach was steeper **c**. This was confirmed by comparing the mean gradients for the two beaches. The mean gradient for the protected beach was 1.57 degrees, compared with 2.01 degrees for the unprotected beach. However, the standard deviation, which provides a measure of dispersion around the mean, was higher for the protected beach. This indicated that the protected beach had a less regular and uniform gradient than the unprotected beach **c**.

The final stage of analysis involved calculation of the Mann-Whitney U statistic. This analysis was very useful because it allowed us to determine whether the differences in profile were statistically significant or merely down to chance **c**. The data were arranged in two columns (one for the protected beach profile and one for the unprotected profile) and their rank order calculated (in relation to both data sets). After summing the ranks, these values were then used in the formula for the calculation of the U statistic. The U value was 89. However, the U tables showed that this value was higher than the critical value at the 95% level and therefore was not statistically significant. In other words, the difference between the two beach profiles could have occurred by chance. As a result we rejected the hypothesis c.

Our analysis enabled us to the conclusion that the two beach profiles were not significantly different. However, there was evidence from the profiles and the dispersion diagram we plotted, as well as the average gradients we drew, that the protected beach was longer but less steep than the unprotected beach d. With a larger number of transects, these differences might have been statistically significant. The analysis was valuable because it prevented us from drawing inaccurate conclusions and it also highlighted possible differences that we did not expect d.

We assumed that the groynes would encourage deposition of sand and shingle and create a steeper beach, but the evidence suggests that this may not be the case. Possible explanations that could be investigated in future include: (a) the unprotected beach, being adjacent to the protected beach, might actually be sheltered by the groynes and experience more rapid deposition; (b) the beach confined between two sets of groynes might experience more scour and wave erosion than the unprotected beach e.

⊕ **17/20 marks awarded**   **a** The context and rationale for the fieldwork investigation are clearly stated at the start of the essay. **b** This paragraph on data collection provides useful background detail but is not strictly relevant to the question. **c** Clear and detailed description and explanation of an impressive range of analytical techniques. **d** Some evaluation, though rather limited in detail and extent. **e** An interesting postscript, but not essential to the answer. Time could have been better spent on providing more evaluation of analytical techniques. The answer shows clear and detailed knowledge and understanding of a range of appropriate analytical techniques. These are applied appropriately in the context of an actual fieldwork investigation. There is some limited evaluation of the techniques and some attempt to show how they were useful to the investigation. The account is well structured with accurate grammar and spelling, and good use of geographical terminology. Sketch diagrams showing the fieldwork site, the beach profiles and dispersion diagrams would help to give a more complete picture of the analysis. Overall, this is a solid, grade-A standard, Level 3 answer. With more detailed evaluation it might have achieved full marks.

Question 7 Evaluation in geographical investigation

Outline the aim(s) of an investigation (fieldwork/research) conducted during your A-level studies. Discuss critically the extent to which the conclusion to your investigation supported its aims.

ⓔ Where appropriate, sketch maps, diagrams and tables should be included in extended-answer questions. Illustrations of this type can help to clarify and demonstrate understanding of sampling strategies, data presentation, data analysis and other techniques used in geographical investigations.

Student A

The fieldwork investigation I am going to write about was based on a small stream called Cam Brook in the Forest of Bowland in Lancashire (see sketch map).

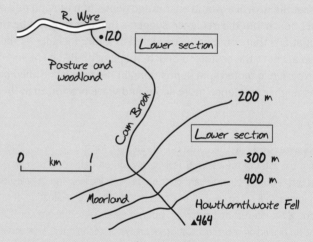

Sketch map of Cam Brook

The stream is a tributary of the River Wyre. It is around 4 km long and its source is 450 m above sea level on Hawthornthwaite Fell. The upper section of Cam Brook is very steep and drains an area of gritstone moorland covered with heather and bracken. However, the lower section (around 1.5 km) has a much gentler gradient, and the valley is well wooded and surrounded by permanent pasture land. There is a thick covering of boulder clay in this area. a

The aim of the investigation was to show that the stream channel should increase in width and depth downstream. The reasoning behind this assumption was that with increasing distance downstream the channel would transport larger amounts of water and sediment. To allow this, the channel would adjust by becoming wider and deeper. b

We collected data at 10 sites and at equal distances along the length of Cam Brook from its source to its confluence with the River Wyre. At each site the bankfull width of the channel was measured. In addition, five measurements of bankfull depth were taken at equal intervals across the channel. Later they were averaged to give a single bankfull depth measurement for each site. We then tabulated the data and analysed the relationship between distance and bankfull width, and distance and bankfull depth using Spearman's Rank correlation coefficient. The correlation coefficient results were

+0.35 for bankfull width and +0.23 for bankfull depth. Although the results showed a weak tendency for both width and depth to increase downstream, which supports the aim of the investigation, unfortunately neither result was statistically significant. This meant that we could not be sure that the results had not occurred by chance c.

Overall this was disappointing. It was not the outcome that I had expected and it required some explanation c. A number of factors could account for the results. The main ones are sample size, accuracy of measurement and other influences on channel width and depth apart from increasing discharge and sediment load with distance downstream d.

Firstly the sample size was quite small. Doubling the number of sites to 20 might have produced better results, even though the value of the correlation coefficients might not have changed very much. Before starting fieldwork we should have had an idea of the size of sample needed to make a correlation coefficient of say 0.4 or 0.5 significant d. Secondly there could also have been some error in the measurements of bankfull width and bankfull depth. At some sites it was hard to determine where the bankfull level was. This was a particular problem where the banks were unequal in height (e.g. on meanders — see diagram) or where dense vegetation (e.g. rushes and reeds) covered the banks d.

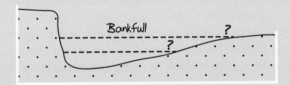

Defining the bankfull channel on a meander

Recording bankfull depth was also difficult in the upper section where large gritstone boulders often clogged the channel (see diagram) d.

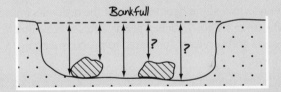

Problems of measuring bankfull depth in boulder strewn channels

Finally, it was clear from our observations that several other factors apart from discharge and sediment load affect channel width and depth. In places in the steep upper section, the channel exposed solid rock (see diagram).

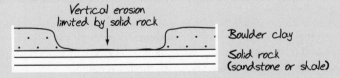

Solid rock and channel shape

This tended to limit vertical erosion and made the channel shallower and wider than expected. In the lower section, the stream was able to cut into soft boulder clay and

the channel tended to be deeper d. Also in this section, the river banks were lined with trees which slowed down sideways (lateral) erosion. In the last 500 m before the confluence with the River Wyre, Cam Brook flowed through fields used for cattle grazing. In places the cattle had trampled the banks, altering the cross-sectional shape of the channel and increasing its width d.

In conclusion the investigation did not succeed in its aim to show that channel width and depth increases downstream. However, it is possible to explain the outcome in terms of (1) the amount and quality of data collected, and (2) the complicated nature of actual drainage basins, where lots of different factors work together to influence the shape of river channels.

⊜ 19/20 marks awarded   **a** A brief description of the location and main physical characteristics of the study area is a good way to begin the answer. **b** A clear statement of the aim and rationale of the investigation is made. **c** Although the results did not support the theory, this does not invalidate the investigation. Rather, it opens up other areas for discussion. **d** A clear and detailed discussion of outcomes and alternative explanations is provided, supported by very specific examples. The answer is based on an actual fieldwork investigation. It has excellent detail and shows good understanding. The answer conveys a critical appreciation of the complexity of fieldwork enquiry and the reasons why there is often a mismatch between theory and reality. The answer benefits from the inclusion of a sketch map, and various sketch diagrams. It is also well structured, and uses accurate grammar, spelling and appropriate terminology. Overall this is an impressive, grade-A standard account, which thoroughly deserves Level 3 attainment.

Student B

Alnwick is a busy market town in Northumberland. It provides goods and services for its residents and for people living in the surrounding area in smaller settlements like Alnmouth, Warkworth and Shilbottle. The aim of my investigation (based on a fieldwork enquiry) was to test the hypothesis that the residents of Alnwick were more likely to shop in the town for convenience goods and services, and less likely to shop for comparison goods and services, compared to people shopping in Alnwick who lived outside the town. **a**

We collected data by interviewing shoppers in Alnwick on a market day (Thursday). We asked shoppers where they lived and whether they had purchased convenience goods and services, and comparison goods and services. We approached people who were shopping, explained what we were doing and asked them politely if they would like to answer our questionnaire survey. Many shoppers declined and said they were too busy. Also as the weather was grey and drizzly many people were reluctant to stop. However, after a couple of hours we managed to get 30 questionnaires completed. **b**

Next we began testing our hypothesis by finding out what proportion of shoppers from Alnwick and from surrounding places bought convenience and comparison goods and services. The results are shown in the table below.

Types of goods and services purchased

	Number of shoppers	Number buying convenience goods/services	Number buying comparison goods/services
Alnwick residents	21	15	6
Shoppers from elsewhere	9	5	4

The results showed that 71% of Alnwick residents bought convenience goods and services compared with 56% of shoppers from outside Alnwick. Meanwhile, 29% of Alnwick residents bought comparison goods and services, compared with 44% from outside Alnwick. These results prove that we achieved our aim and that people who live in a market town are more likely to be shopping for convenience goods and services than people living elsewhere. They are also less likely to shop for comparison goods and services. c

While the investigation proved to be successful it could have been improved by collecting more data and increasing the size of the sample d. This could be done by spending 4 hours rather than 2 on interviews. Also, some shoppers didn't understand what was meant by 'convenience' and 'comparison' goods and services, and these could have been explained more fully, though we only had a limited amount of time available d.

e 10/20 marks awarded   a This is an authentic investigation based on primary fieldwork data. The location and aims of the enquiry are presented in the opening paragraph. b The answer often loses sight of its objective, with a good deal of irrelevant description on data collection and associated problems. c The results of the investigation are analysed rather uncritically with no analysis of statistical significance. Despite the conclusions in the fourth paragraph, it is likely that the outcome of the enquiry did not, in fact, verify the hypothesis being tested. d Some implied evaluation is found in the concluding paragraph but, overall, critical understanding of the investigation and its outcomes are lacking. This answer is superficial and shows a naive understanding of the investigation. There is much irrelevance; conclusions are not always supported by the evidence presented; and the quality of the primary data collected used in the analysis is dubious. Evaluation and critical discussion are limited. On the positive side the answer is built around an actual investigation. It has a sound structure, no obvious errors of grammar and spelling and uses some appropriate geographical terminology. Yet, overall, this is a modest grade-D standard answer, which achieves only the top end of Level 1.

Knowledge check answers

1 Some hypotheses are impossible to investigate because primary documentary data do not exist/are unavailable; access to fieldwork sites is restricted; timescales may be inappropriate (e.g. study of bedload movement, floods).

2 A hypothesis is a statement of difference or relationship between variables that is testable using scientific methodology. In contrast, a theory is a set of statements or principles that have been tested and proved valid.

3 Likely target populations are: shoppers in a city centre; commuters in a residential area; suspended sediment load and bedload; cross-sections of river channels.

4 Factors that must be taken into account to minimise risk and ensure personal safety in geographical fieldwork include: location (e.g. isolation, accessibility) and how easy it would be to get help if problems arose; potential hazards associated with working in physical and human environments (e.g. rivers, coasts, mountains, cities); and responses to these hazards such as working in groups, wearing appropriate clothing, headwear and footwear, checking weather forecasts, checking tide tables, carrying emergency equipment (e.g. survival bag, torch, whistle), mobile phones, leaving itinerary details with a responsible adult, politeness and sensitivity towards members of the public, avoiding urban locations where perceived risks are high (e.g. areas of urban dereliction or of severe multiple deprivation).

5 It is important that care is taken when using equipment. Losing a pair of callipers or breaking a clinometer during the morning of a day of investigation places at risk any further reliable and accurate data collection.

6 Spatial sampling involves the collection of data either at points, in areas (quadrats) or along lines (transects). Samples are then selected either randomly or systematically. Scree particles on a talus slope could be sampled at regular intervals (i.e. points) along random transects established perpendicular to the base of the slope. In vegetation studies quadrats could be used to record sample data on the number of plant species.

7 Possible criteria include: clarity; visual effectiveness; accuracy; detail; potential for accurate data retrieval.

8 Population is a discrete variable (i.e. located at specific points). Isopleth maps are most effective for representing data distributed continuously in space (e.g. pressure, rainfall). Isopleths, drawn by interpolation, assume regular change in a variable over distance (not the case with population). There are problems of accuracy using isopleths where population is (a) low density or (b) very high density. Isopleths are highly generalised and fail to show variable spatial patterns in detail.

9 Choropleth maps are relatively easy to draw and bundles of population data are readily available for areal units such as wards and census super output areas. However, abrupt changes of density (which do not reflect actual discontinuities in population distribution) often occur at areal unit boundaries. Moreover, these areal units can vary substantially in size and shape, strongly influencing the appearance of the map. Choropleth maps provide no information on the internal distribution of population with areal units — a particular problem where these units cover large areas. Data can be extracted from choropleth maps for individual areas, though in this example it comprises just broad classes of population density.

Within cities, population distribution is approximately continuous. Isopleth maps are effective in representing such distributions (other examples of continuous distributions are rainfall, pressure and temperature). In urban areas which are relatively small, detailed isopleth maps (comprising up to 10 isopleths) would be too crowded and too complex to draw and to interpret. But using fewer isopleth means that maps become excessively generalised. Also, sudden changes in population density within urban areas (e.g. inner city, suburbs, rural–urban fringe) are difficult to accommodate on isopleth maps, causing further generalisation. For this reason extracting accurate data from isopleth maps is problematic. Nonetheless, isopleth maps can provide a good (albeit generalised) overview of population distribution in medium-sized cities.

10 The median is most useful when data sets are skewed (i.e. have non-normal distributions). The mean is not representative of skewed data sets: its value is strongly influenced by extreme values.

11 Mann-Whitney U: environmental quality is higher in neighbourhood A than in neighbourhood B.

Correlation: the length of the growing season for crops is influenced by altitude; rent values decrease with distance from a city centre; steam discharge increases with drainage basin area.

12 Length of the growing season for crops (y), altitude (x); rent values (y), distance (x); steam discharge (y), drainage basin area (x).

13 Statistical significance is important to determine the likelihood that the outcomes based on samples have occurred by chance. Without determining the statistical significance of differences or relationships, we can have no confidence in the results of geographical investigations.

14 The U test compares two data sets to determine whether the difference between them is statistically significant. It could be used, for example, to compare: distances travelled by shoppers to district and neighbourhood shopping centres; slope angles in chalk and clay landscapes; and size of limestone and sandstone particles in a river channel (or on a shingle beach). Chi-squared is used to compare differences between frequency distributions. Examples include: the distribution of spending by shoppers in different shopping streets/shopping centres in a CBD; differences in the slope profiles of sand and shingle beaches; and differences in the lengths of journeys by visitors to national parks and country parks. Spearman Rank correlation is a test of association between two variables (where x is an independent variable and y is a dependent variable). Examples are: drainage basin area (x) and river discharge (y); journey times (x) and the volume of commuting (y) to a major employment centre; and altitude (x) and rainfall (y).

15 Among possible methodological reasons for this unexpected outcome are: samples of shoppers too small to give statistically significant results; unrepresentative samples because many shoppers declined interviews; timing of questionnaire surveys may have been different (i.e. time of day, day of week); and failure of the survey to differentiate shoppers arriving by car and those arriving by other forms of transport or on foot (i.e. stratification of the population).

OCR A2 Geography